FIELD TESTING OF WATER IN DEVELOPING COUNTRIES

by

L G HUTTON

WEDC
Department of Civil Engineering
University of Technology
Loughborough
LE11 3TU

First published in 1983 by
Water Research Centre
Medmenham Laboratory
Henley Road
Medmenham, PO Box 16
Marlow, Bucks SL7 2HD
England
© 1983 Water Research Centre and L. G. Hutton

ISBN 0 902156 06 3

DISCLAIMER

The mention of specific companies or of certain manufacturer's products does not imply that they are endorsed or recommended by the Water Research Centre in preference to others of a similar nature that are not mentioned. Errors and omissions excepted, the names of proprietary products are distinguished by initial capital letters.

The author wishes to thank those companies mentioned for assistance in preparation of this publication and other firms who of course market similar products are asked to contact the author to ensure they are not overlooked in future works. The field of water testing is expanding continuously and users of equipment all over the world are also asked for their comments on the performance of apparatus whether included or not in this publication.

Printed in Great Britain by
Unwin Brothers Limited,
The Gresham Press, Old Woking, Surrey

Preface

V. K. Collinge, Director of Planning,
Water Research Centre

The Water Research Centre (WRC) is the research arm of the British Water Industry. It undertakes research on most aspects of water technology and has a staff of about 500. The Centre has three laboratories dealing with, respectively, processes for water and wastewater treatment, environmental studies, and the engineering problems of underground pipes.

WRC has carried out many studies overseas, often in support of British Consultants or for International Organizations such as the World Health Organization (for which it is the Collaborating Centre for Drinking Water and Water Pollution Control).

The International Drinking Water and Sanitation Decade poses a tremendous challenge with its central objective of providing access to safe water and adequate sanitation for all by 1990. In response to this, the WRC is increasing its involvement with the water and sanitation problems of the Developing Countries and aims to deploy its skills in support of the Decade.

WRC has had wide experience of the testing of water quality, but mainly confined to modern laboratories. The need to carry out tests of water quality in the field has led to the production of a number of portable test kits. However, to date, no comprehensive 'state of the art' review of these kits has been published, and a need clearly existed for such a review.

This book, written by Mr L G Hutton, of Water and Waste Engineering for Developing Countries Group (WEDC) of Loughborough University of Technology and published by the Water Research Centre, is designed to fill this need.

July 1983

Contents

1. Introduction

The main objective of this report is to review field methods for testing water in developing countries. At present there is no book available that covers a range of possible techniques, although several companies publish guides to their own products. Laboratories in developing countries are often poorly equipped and lack financial and staff resources to enable them to cope with the ever increasing demands on their services.

It is hoped therefore that this publication will be used to provide ideas for suitable equipment, taking into account the approximate cost and sources of materials.

It is also anticipated that consultants and advisers may find this publication of value to help them select and use equipment for rapid field assessment of water quality.

It will be of use to engineers, scientists (such as chemists, geologists and bacteriologists) and laboratory technicians testing quality of water sources and treated water.

Particular reference is given to bacteriological analysis, since the detection of faecal pollution of water is very critical in developing countries where pathogenic-bacteria-related diseases are rife. Diseases such as cholera and typhoid receive publicity but the biggest killers are diarrhoea, gastroenteritis and dysentery, especially among young children.

Many laboratories analysing water overseas are staffed by chemists whose training is very much 'Western oriented' and unlikely to have included much bacteriology. Most bacteriological techniques have been self-taught by chemists. Conversely the few bacteriologists need some chemical background if they are to progress to higher management positions in charge of water testing facilities. This publication will serve as an introduction and guide to the available literature for both groups.

When selecting a particular piece of equipment for field water testing the following questions should be asked:

1. Will it cover the range of concentrations expected?
2. Can we afford to buy it?
3. Is it suitable for our purpose?
4. Can it be used for other determinations?

Recent developments in equipment for field testing encourage the use of basic equipment for several applications. For example, simple colour comparators can be used with different discs and reagents: specific ion meters can be used for fluoride, pH, nitrate, and dissolved oxygen determinations and the HACH digital titrator enables several titrimetric procedures to be undertaken in the field such as calcium, magnesium, alkalinity, free CO_2, dissolved oxygen, chloride, chlorine etc. The efficient use of basic equipment enables the field worker to enlarge his repertoire of determinations gradually as finances permit or demands require.

The nature of a particular parameter may require tests to be made in the field because its value may change rapidly between sampling and analysis in a laboratory perhaps days (or even weeks) later. The advantage of field tests is that they give rapid results that can be repeated on the spot if necessary. Parameters such as nitrate, nitrite, chlorine, dissolved oxygen, pH, Eh, alkalinity, iron and manganese all change rapidly after sampling and values are more relevant if determined in the field.

If pollution is detected by field testing it is sometimes possible to determine the possible sources immediately and even to take remedial action. Delays cost money and may even cost lives.

When doing field tests it is important to involve the local community. Tell them why and how you are testing the water. After all it is *their* water supply.

Another often compelling reason for field testing is the absence of any efficient water testing facilities in the region.

Cheaper equipment is needed especially for bacteriological analysis, but being pragmatic you only get what you pay for.

It is important to regard this publication as a discussion document and merely as a starting point. Field workers overseas are invited to comment on the performance of field testing kits for water whether mentioned in this publication or not. The author will welcome such information. Please write to the following address:

Mr L G Hutton
WEDC Group
Department of Civil Engineering
University of Technology
Loughborough, Leics LE11 3TU
England.

1.1 Guidelines for water quality

In proposing methods for water analysis there is a need to consider how the results will be applied.

The relationship between water quality and health effects has been studied for many water quality characteristics. An examination of water quality is basically a determination of the organisms, and the mineral and organic compounds contained in the water.

2

There is a wide range of test kits available and it is suggested that certain critical minimal tests be carried out on drinking water. These are in order of importance:

1. Faecal coliform organism count
2. Turbidity
3. Chlorine
4. Conductivity
5. pH

Details of why these tests are critical are given in the text.
The basic requirements for drinking water are that it should be:

● Free from pathogenic (disease causing) organisms.
● Containing no compounds that have an adverse effect, acute or in the long term, on human health.
● Fairly clear (i.e. low turbidity, little colour).
● Not saline (salty).
● Containing no compounds that cause an offensive taste or smell.
● Not causing corrosion or encrustation of the water supply system nor staining clothes washed in it or food cooked in it.

For their ready application in engineering practice, the results of the studies and research on drinking water quality must be laid down in practical guidelines. (Huisman *et al.*, 1981.)

The World Health Organization (WHO) have recently published their WHO Guidelines for Drinking Water Quality (WHO, 1983) in a three-volume format and it is hoped that they will be read in conjunction with this publication. The contents of the guidelines are as follows: Volume I will present the recommended guidelines values *per se* together with essential information required to understand the basis for the recommended guideline values as well as information on monitoring requirements and where possible suggestions regarding remedial measures. This volume is primarily intended for those who are engaged in the standard setting process, as well as those that are responsible for the provision of safe drinking water. For each of the recommended guideline values, the toxicological and epidemiological basis for choosing a given value, and the health risk involved, are summarized including information on uncertainties, safety factors, multimedia routes or exposure, etc. Special attention is also given to the ways and means in which these guideline values are to be applied and used.

The second volume, which will contain some 400 pages, is essentially an environmental health criteria document covering those substances/contaminants which were examined with a view to recommending guideline values. This volume contains a review of the toxicological, epidemiological and clinical evidence which is available. This volume elaborates greatly on the health risk information presented in Volume I and should be considered as a vital companion document. Volume III of these Guidelines is intended to serve a very different purpose. It will contain recommendations and information concerning what needs to be done in small communities and in rural areas with respect to safeguarding their water supply. While it will detail some methodology concerning sampling and analysis of water supplies, a much greater emphasis is being placed on sanitary surveys and similar means of investigating the possibilities of contamination, particularly due to the presence of pathogenic organisms. The sampling and analysis of water supplies are limited to the basic techniques of multiple tube and membrane filtration as concerns bacteriology and simple methods for residual chlorine determination. This third volume is intended primarily for those authorities and people at the community level whose responsibilities entail the protection of public health or who may work in areas of general sanitation, etc. This volume is to be produced in much greater number and in more languages than will be possible with the other two in the hopes that these Guidelines can reach and be used by the local authorities in as many developing countries as possible.' (Gorchev, H. G. and Ozokins, G., 1982).

The Guideline values for the chemical parameters are included in the relevant sections of Chapter 3 of this publication on field methods, but it is important that the advisory nature of the Guideline values be recognised, and that they are used in evaluating water quality in the manner described in the new WHO publications and not in isolation.

2. Bacteriological Analysis

The main purpose of bacteriological examination is the detection of recent, and therefore potentially dangerous, faecal pollution. Contamination of drinking water by human or animal excrement, or by sewage, is dangerous if among the contributing population there are cases or carriers of infectious enteric diseases which may be waterborne. Many different pathogens (disease-causing organisms) could be present in faecally polluted water and it is not practically possible to test for them all in a particular sample. The testing procedure for pathogens is complex and takes several days. The pathogens themselves may only be present occasionally in the water although pollution by faecal material (excreta or sewage) might be occurring continuously. We do not determine whether pathogens are actually present but we determine if the water is polluted by faecal material. Consequently one looks for the presence or absence of types of bacteria which are ALWAYS present in the excreta and intestines of man and other warm blooded mammals (such as cows, sheep, pigs, poultry etc.) and whose presence in water therefore INDICATES faecal pollution.

There are three main groups of INDICATOR organisms which can be used to detect FAECAL pollution:

1. The COLIFORM group, deriving its name from the colon or large intestine and typified by *Escherichia coli (E.coli)*.
2. The FAECAL STREPTOCOCCI group typified by *Streptococcus faecalis*.
3. The spores of the SULPHITE REDUCING CLOSTRIDIA group typified by *Clostridium perfringens*.

The presence of such organisms INDICATES faecal pollution and thus that intestinal pathogens could be present. Conversely the absence of the faecal INDICATOR organisms indicates that pathogens are probably also absent. See Figure 2.1.

FAECAL INDICATOR ORGANISMS

ABSENT   PRESENT

PATHOGENS   PATHOGENS
PROBABLY   MAY BE
ABSENT   PRESENT

Figure 2.1
Rationale for
use of indicator
organisms

The search for such indicators of faecal pollution provides a means of assessing the hygienic quality of drinking water whilst the surveillance of the bacteriological quality of raw water is important in assessing the degree of pollution, and choice of the best sources and their treatment. (WHO, 1983.)

In this report the analysis for coliform group of organisms above will be described because:

1. They are the most important group of bacteria in sanitary engineering and meet most of the criteria of the 'ideal' indicator organism.
2. They can be determined under field conditions.
3. The testing procedures for streptococci involve longer incubation periods and often the use of agar which is difficult under field conditions.
4. Clostridia require heating to 80 °C and a double fermentation procedure (WHO, 1983) which would be very difficult to perform under field conditions.

The reader is referred to WHO (1983), APHA (1980), HMSO (1969), and Mara (1974) for details of laboratory procedures for the detection of faecal streptococci and sulphite-reducing clostridia. These other faecal indicator organisms do have a role to play in confirming the origin and nature of excremental pollution. Their procedures, however, are not easy under field conditions.

2.1   Coliform group of organisms
The coliform group of organisms are used worldwide as indicators of faecal pollution being normally associated with faeces and water. They are characterised broadly by their ability to ferment lactose in culture at 35 °C or 37 °C with the production of both acid, aldehyde and gas within 48 hours. However, there are other sources of coliform organisms apart from the faeces of warm-blooded animals, namely vegetation and soil.

Certain coliform organisms retain the ability to ferment lactose at 44 °C or 44.5 °C. These are faecal coliform organisms which may sometimes be described as thermotolerant coliform organisms. These include the genus *Escherichia* and to a lesser extent occasional strains of *Enterobacter, Citrobacter* and *Klebsiella*. Of these organisms only *Escherichia coli* (or *E. coli* for short) is specifically of faecal origin being always present in the faeces of man, animals and birds in large numbers and is rarely found in water or soil which has not been subject to faecal pollution. The presence of *E. coli* is regarded as definite proof of faecal pollution.

In temperate climates the faecal coliform organisms detected at 44 °C or 44.5 °C consist predominantly of *E. coli*, but in hot climates less than 50% of the faecal coliform organisms are probably *E. coli.* The absence of faecal coliform organisms, especially when (total) coliform organisms have been detected, does not necessarily rule out the possibility that faecal contamination may have occurred. In such cases, a search for faecal streptococci and clostridia could be undertaken to confirm the faecal origin of pollution. However, as indicated above they are beyond the scope of this book and in most cases the absence of faecal coliform organisms in a water implies the water was not faecally polluted at the time of the sampling.

Therefore, when interpreting the results of the bacteriological examination of water, these points must be taken into account. So determinations of 'Total' coliform organisms at 35 °C or 37 °C which result in the detection of a small number of coliform organisms (1-10 organisms per 100 ml) may be of limited sanitary significance if faecal coliform organisms (see below) are absent also (WHO,1983). In treated and distributed water supplies the presence of coliform

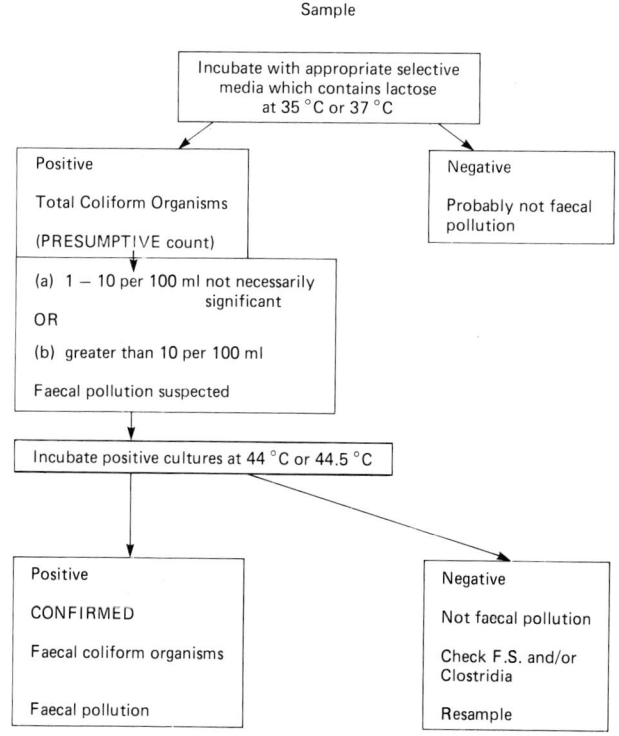

Figure 2.2 Simple flowsheet for coliform organism determination

organisms is unacceptable especially if these organisms are shown to be faecal coliform organisms. Similarly, in untreated water supplies whether distributed or not the detection of coliform organisms, especially faecal coliform organisms, is undesirable since there is some evidence that pollution has occurred. Rural community water supplies in developing countries rarely at present meet the WHO (1983) bacteriological Guidelines for the detection of coliform organisms. Certain coliform organisms grow in soil in warm countries and may be washed into the water supply. Therefore low levels of these particular organisms do not necessarily indicate faecal pollution. Excessive numbers of coliform organisms and any detection of faecal coliform organisms in water should always be viewed with suspicion. In the long term the objective is to eliminate the use of faecally polluted sources but this objective must be tackled pragmatically.

The Total coliform count is a presumptive test for faecal pollution. In order to confirm that they are of faecal origin a confirmatory test with incubation at 44 °C or 44.5 °C is required.

Since the primary purpose of bacteriological analysis is to detect faecal pollution the incubations at 44 °C or 44.5 °C provide the most direct route to detecting faecal coliform organisms (see Figure 2.3).

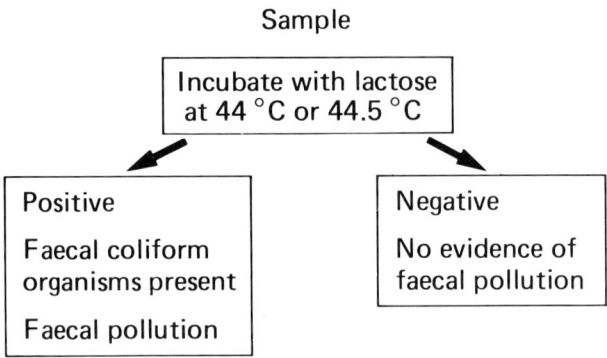

Figure 2.3   Faecal coliform organism determination

2.2   Aseptic procedures

Since we are looking for very low levels of indicator bacteria, perhaps 1 or 2 per 100 ml of sample, the sample or apparatus must not be contaminated by careless handling. Man excretes 13×10^6 faecal coliform organisms per gram of excreta per day and it would be foolish to contaminate a sample merely because hands were not washed after going the the toilet. Similarly, bearing in mind the faecal-oral route of transmission of disease (Figure 2.4) the working area should be free of dust, flies and food.

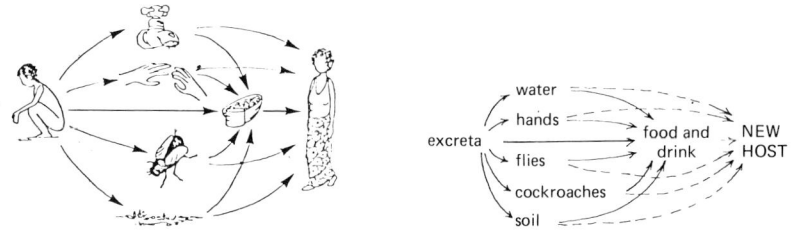

Figure 2.4 Faecal-oral route of disease transmission

So before we begin analysis we should:

1. Clean the working area.
2. Wash our hands.
3. Avoid dusts and draughts in the working area.
4. Use only media and equipment known to be sterile.
5. Avoid touching any part of a container, petri dish, pipette, etc., which will come in contact with our sample.
6. Only open sterile packs or petri dishes for as short a time as possible.
7. Keep the round tipped forceps immersed in at least 1″ of alcohol (methyl or ethyl) and rapidly burn off the alcohol before using them.
8. For field work use disinfecting cloths with self-indicating strips both for preparing the working area and cleaning hands before and after testing. They are made by Wipex Labpak, and can be obtained from: Baird and Tatlock Ltd. 100 pack costs £28.00.

A portable butane or gas torch is also useful for field work, especially for sterilizing taps and stainless steel apparatus — Camping Gaz makes a very handy torch. Remember that gas cylinders cannot be transported by air so look for local sources.

2.3 Sampling storage and transport of samples for bacteriological analysis

There is a saying that 'the analysis is only as good as the sample'. The sample should be representative of the bulk of the water being examined and care should be taken to ensure that there is no accidental contamination of the sample during sampling. Sample collectors should be trained and made aware of their responsibilities. The samples should be

9

clearly labelled with the site, date, time, nature of the water and either analysed on the spot with the field kits, or sent to a suitable laboratory for analysis without delay in a suitable transport and storage container. Samples which cannot be analysed within 24 hours of sampling could be membrane filtered on site and transferred to transport media before final examination and incubation within 3 days. (WHO, 1983 and AWWA, 1980.)

2.3.1 Preparation of sample bottles

Glass bottles of at least 200 ml capacity with a ground-glass stopper or rubber-lined aluminium or plastic screw caps should be cleaned and washed thoroughly, then rinsed with distilled or deionised water. Polypropylene bottles which will withstand autoclaving may also be used.

If the water to be examined is likely to contain chlorine or other disinfectant then 0.1 ml of 1.8% solution of sodium thiosulphate ($Na_2S_2O_3.5H_2O$) per 100 ml of bottle capacity should be added to neutralise any residual disinfectant (up to 5 mg/l available chlorine). The addition of thiosulphate at this concentration has no significant effect on the faecal coliform organisms whether the water is chlorinated or not. (HMSO, 1969.)

For ground-glass-stoppered bottles only, a small strip of clean paper should be placed in the neck of the bottle with enough protruding to allow its removal and thereby ease the top from the bottle prior to sampling.

A piece of paper, preferably Kraft paper or aluminium foil (although any thick paper, even newspaper may be used) should be fastened over the top and neck of the bottle and tied with string (Figure 2.5). (It should be an open bow knot which can be released by a single pull.)

Aluminium test-tube caps can also be stuck to the tops to protect the neck of the bottle. The screw top should only be loosely fastened prior to sterilizing and then only tightened when cooled following sterilization.

2.3.2 Sterilization of sample bottles

The sample bottles can then be sterilized (i.e. all micro-organisms on the glass or in the bottles killed) by 'autoclaving' in a pressure cooker or by heating them in a dry oven at 170 °C for 1 hour.

The autoclaving can be carried out in a domestic pressure cooker such as the Prestige 'Hidome' (cost £20-£25) provided it can withstand and operate at 15 lbs per square inch gauge pressure thus producing steam at a temperature of 122 °C for 20 minutes. With the Prestige 'Hidome' model this means using all the weights together. If operating the pressure cookers over 600 metres above sea level in-

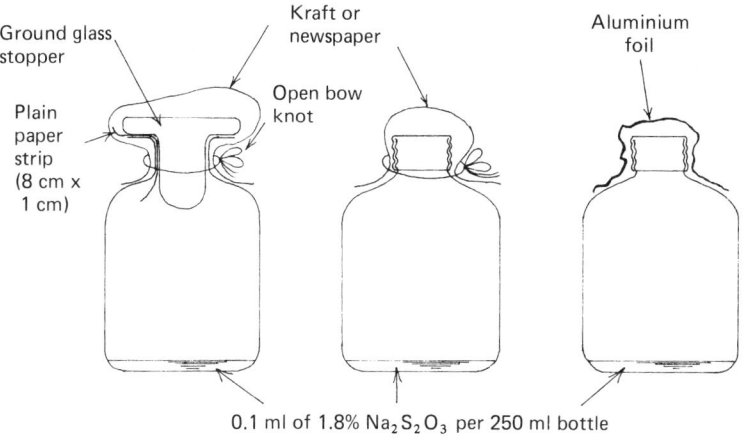

Ground glass stopper

Plain paper strip (8 cm x 1 cm)

Kraft or newspaper

Open bow knot

Aluminium foil

0.1 ml of 1.8% $Na_2S_2O_3$ per 250 ml bottle

Figure 2.5   Sterile bottles and protected tops

crease the usual cooking or sterilizing time from 20 minutes by 1 minute for every 300 metres (1000 ft) above sea level.

The pressure cooker can also be used to sterilize other materials used in bacteriological work such as dilution bottles, small cylinders, beakers, tubes, buffer solutions and small pipettes.

2.3.3   Sampling procedures

When several samples are being taken for water quality testing at the same location, the sample for bacteriological examination should be collected first to avoid danger of contamination of the sampling point.

Samples for bacteriological analysis must only be collected in sterile bottles. The bottle should be kept unopened until it is to be filled. The stopper should be removed and during sampling neither it nor the neck of the bottle should be allowed to touch anything. The bottle is best held by the other hand around the base of the bottle. Do NOT rinse out the bottle with the sample and do not completely fill the bottle. This will allow the sample bottle to be shaken prior to analysis.

In some cases it may not be possible to collect the sample directly in a sample bottle, and a sterile stainless steel jug or cup should be used. The stainless steel container can be sterilized by igniting methyl or ethyl alcohol inside it. The water collected should be transferred to the sample bottle.

Samples are taken to be representative of the bulk of the water under examination. Surface waters such as rivers, lakes, streams, reservoirs, should be sampled away from the banks if possible. In tropical areas the sampler is putting himself at risk from schistosomiasis (bilharziasis)

11

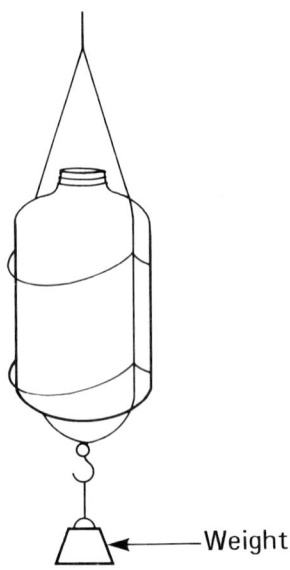

Figure 2.6   Sampler for shallow wells

when he is in the water and he should therefore wear waterproof gauntlets and long boots to prevent the cercariae from penetrating the skin and invading his body. The outside of the sample bottle also needs to be carefully rinsed with disinfectant to kill off any cercariae on the bottle. A Wipex cloth could be used to wipe the bottle. In tropical regions the medical authorities should be consulted about the risk of schistosomiasis and for advice before sampling takes place.

It should be remembered that the numbers of bacteria being determined can be very small and the sample being collected should not have any contact with the person, clothing or any object used to assist the collector in approaching the sample site.

If by accident the sample bottle thread or the inside of the cap or stopper is touched the sample bottle should be discarded and another sterile bottle should be used. It is good practice to have some spare sterile sample bottles on your person when sampling.

2.3.4   Sampling from surface waters

If the sample is being taken from a source of supply make the collection as close as possible to the intake to the treatment plant. Samples should be taken about 30 cm (1 foot) below the surface or at mid-depth

12

if the depth of the stream is less than 30 cm. Hold the bottle by the base with the opening into the current.

When full bring the bottle to the surface and CAREFULLY replace the cap on the bottle.

2.3.5 Sampling from wells or boreholes

For wells or boreholes equipped with a pump, the pump should be operated to clear any standing water in the water column (this pumping could be at least 20-30 minutes depending on the depth and diameter of the borehole) before the outlet pipe is sterilized using either a gas

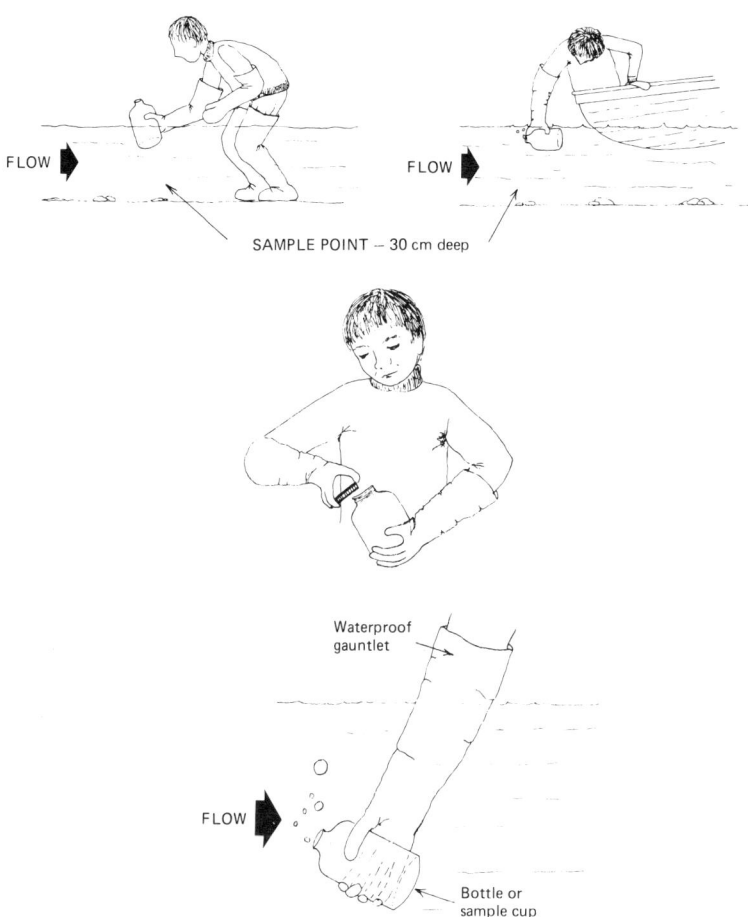

Figure 2.7 Sampling from streams

torch or alcohol flaming. Operate the pump for a further 2 minutes and take a sample in the flowing stream of water. For shallow, open wells a weighted bottle or shallow sampling device may be used.

A sample from the water carrier's pot or bucket may be more representative of what is actually being drunk so take a sample poured into the bottle from her bucket as well (see Figure 2.8).

Figure 2.8   Sampling from water carrier's container

If the water is stored in the household then the household container would also need to be sampled. Positive results from the household would indicate bad hygiene in the household rather than polluted groundwater.

2.3.6   Sampling from a tap
First check that the tap is fed directly from the pressure mains and not a cistern or roof tank. Remove any nozzles or filters from the tap and if the tap leaks replace the washer. Clean the outside of the tap. Turn the tap full on and allow it to run to waste for 4-5 minutes. Close the tap, dry it with a clean cloth (Wipex are useful for this) and sterilize the opening of the tap with the gas torch or alcohol flame. Turn on the tap and after 1-2 minutes carefully collect a sample from a flow run at the normal usage rate but avoiding splashing (see Figure 2.9). When sampling from copper or galvanized iron pipes, the flushing time should be increased as the metals can have a bactericidal effect on the sample.

2.3.7   Sampling from a chlorinated supply
If a supply is chlorinated it is necessary to neutralise any residual chlorine in the water otherwise any bacteria present will be killed or prevented from growing on the culture medium. Sodium thiosulphate is used to inactivate the chlorine and is usually added to the sample

14

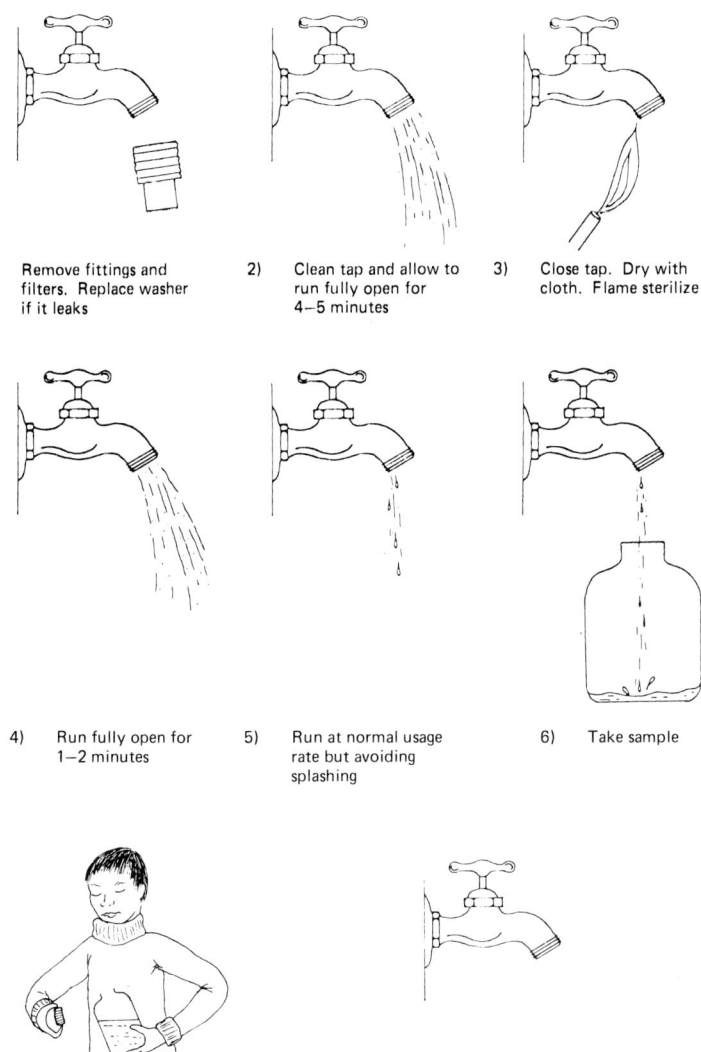

1) Remove fittings and filters. Replace washer if it leaks

2) Clean tap and allow to run fully open for 4–5 minutes

3) Close tap. Dry with cloth. Flame sterilize

4) Run fully open for 1–2 minutes

5) Run at normal usage rate but avoiding splashing

6) Take sample

7) Replace stopper carefully

8) Turn off tap and thank consumer for use of his tap

Figure 2.9   Sampling from taps

bottles in solution before sterilization (see 2.5). It is normal practice to add the sodium thiosulphate to all sample bottles so that the sterile bottles can be used for chlorinated or unchlorinated supplies.

Because sodium thiosulphate has been added sterile sampling bottles are not rinsed with the samples before sample collection for bacteriological analysis.

Determine the concentration of residual chlorine at the sampling point at the time of sampling. (See Section 3.21, Chlorine.)

2.3.8 Transport and storage of sample

If samples are not examined on site they should be stored in covered, cooled insulated containers (cool boxes) during transport so that their temperature is kept between 4 °C and 10 °C but not frozen. The use of ice packs and ice bricks is recommended. Care must be taken to prevent contamination of the sample by the cooling agent.

Examination of the samples should begin as soon as possible and WHO guidelines (1983) suggest within 24 hours provided they are kept cool and dark. The time delay between sampling and examination should be noted and stated to assist intepretation of results.

It may be easier to examine or process the samples when there is a suitable clean place or convenient time to examine them in batches and place them in the incubator together. This is probably done best in the middle of the day and in the evening so that the incubator can emptied and have space available for the morning or afternoon sampling run.

If it is not possible to process and incubate the sample within 24 hours, it should be filtered as normal and placed on an absorbent pad saturated with 'transport medium' and sealed in a petri dish. The transport or holding medium is a dilute broth which allows the bacteria to remain alive but does not permit the formation of visible colonies in three days (HMSO, 1969; APHA, 1980).

As soon as the incubator is available the membrane is transferred aseptically to a pad saturated with the appropriate medium and incubated for the normal times.

The specification for holding medium according to Report 71 (HMSO, 1969) is:

> Peptone 0.2 g
> Sodium chloride 0.1 g and 1 litre water
> Sodium benzoate 4 g

It should have a pH of 7.5. The solution is best prepared fresh and then distributed into smaller bottles for sterilizing before field use.

2.4 Frequency of sampling

This is usually determined by local conditions in developing countries. Ideal frequencies relating to populations are described in several WHO publications (McJunkin, 1976; WHO, 1971; WHO, 1983) but few countries can afford or are able to meet their recommendations due to lack of trained staff and resources. Perhaps the criteria should be:

1. To test as often as possible.
2. To test in as many points in the water network as practicably feasible.
3. To keep the testing facilities fully employed until an acceptable frequency of sampling is obtained.

There should be a concentration of testing at points where the maximum benefit will be obtained.

2.5 Techniques available for field detection of indicator organisms
There are two procedures available for counting the number of faecal indicator organisms present in a sample of water:

1. The multiple tube fermentation technique.
2. The membrane filtration technique.

2.5.1 The multiple tube fermentation technique
The multiple tube fermentation technique involves adding measured volumes of sample to sets of sterile tubes or bottles containing a suitable liquid medium containing lactose. After incubation at 35 °C or 37 °C for an appropriate period, usually up to 48 hours for total coliform organisms (with an inspection at 24 hours), the tubes are checked for positive reactions, in the case of coliform organisms these are acid and gas production. The gas production is detected by its appearance in a Durham tube inserted into the tube or bottle (Figure 2.10).

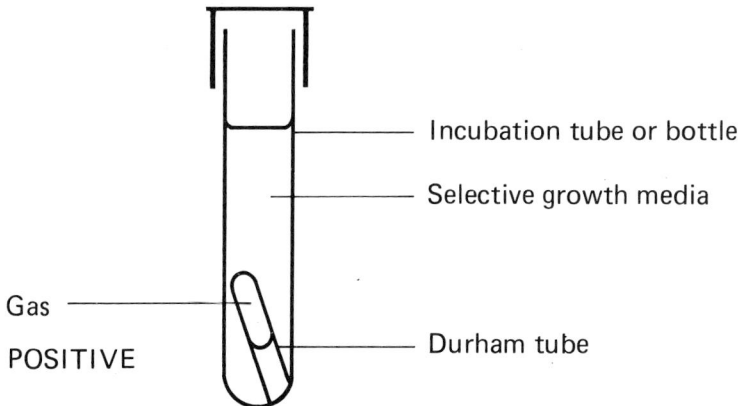

Incubation tube or bottle

Selective growth media

Gas

POSITIVE

Durham tube

Figure 2.10   Gas production detected by Durham tube

Acid is detected using various indicators. The number of tubes showing both positive and negative reactions are recorded at the end of the incubation period and an estimate of the most probable number (MPN) of organisms present in the original sample obtained by using appropriate statistical tables relating to the volumes of sample inoculated. The original sample should be diluted to ensure that both positive and negative reactions are obtained. These tables are found in APHA (1980), WHO (1983), HMSO (1969).

The multiple tube fermentation or MPN technique is applicable to waters of all types and especially those with high turbidity. The equipment required is relatively cheap and simple (positive reactions being easy to read). The multiple tube fermentation, however, gives only a statistical estimate of the most probable number of coliform organisms present.

In order to confirm if these organisms are of faecal origin further tests must be made on samples taken from those tubes showing positive reactions following the first incubation. These confirmatory tests are made by taking a small amount of the positive culture (using a sterile wire loop) and introducing this into tubes of media containing lactose and bile salts for incubation at 44 °C or 44.5 °C for 24 ± 2 hours. The production of gas (detected by its presence in Durham tubes) CONFIRMS the presence of faecal coliform organisms in the original tubes and once again by using statistical tables an estimate of the most probable number (MPN) of faecal coliform organisms can be made.

The problems of using the multiple tube fermentation (or MPN) technique in the field are:

1. Lack of a portable incubator for tubes or bottles able to operate at 35 °C and 44 °C or 44.5 °C.
2. Long time for full confirmation of faecal coliform bacteria (up to 3 days).

Its use is recommended where there is a reliable electrical supply that can be used to power cheap incubators such as the HACH 2078 − 02 (Cost £125) or the GRANT JB2 + LJ2 Water Bath (Cost £129). Research into the production of cheaper incubators is being carried out by Professor Duncan Mara at Leeds University and Dr. Barry Lloyd at the University of Surrey (Personal Communication 1982, 1983). For the latest details the reader should contact them directly. If there is any chance of using a fixed incubator on a reliable power source both the Multiple tube method and the membrane filtration method will be cheaper. The list of factors to be considered when making the choice between techniques is summarised in Table 2.1 and should be used in conjunction with the algorithm in Figure 2.28.

There is an urgent need in developing countries for such an incubator especially if it can be battery powered since the power supplies in the

18

Table 2.1 Comparisons and contrasts between the multiple tube and membrane filtration techniques for field bacteriological testing of water in developing countries.

Multiple Tube MPN	Membrane Filtration
Costs	
Large quantities of culture media and glassware and large autoclave.	Smaller quantities of media. Membrane expensive. Disposables (petri dishes, pipettes) expensive.
Capital costs fairly high.	Capital costs of proprietary equipment high.
Accuracy	
Statistically based estimate. Possibly large errors especially at low levels.	More accurate especially to low levels (less than 100 colonies) (100 ml)
Field use	
Needs static base.	May be operated in the field and in transport if portable incubator used. Transport media can be used prior to incubation.
Suspended matter	
May be used for turbid samples.	Not suitable for turbid waters due to membrane clogging.
Convenience	
Large amount of material to be prepared prior to analysis and disposed of after incubation.	Less manipulation and hence lower chance of contamination. Disposable, pre-sterilized equipment can be purchased.
Incubation times	
Up to 48 hours or 72 hours.	24 hours (or even 7 hours in some special cases, (WHO, 1983).

third world are unreliable. Hach Chemical Company market closed vials which contain selective growth media for faecal coliform organisms costing £2.80 for 5 tubes. The operation of a simple multiple tube technique is described by Mara in Cairncross and Feacham, (1978). The current prices of equipment described by Mara are listed in Table 2.2.

This publication aims to describe the field testing of water in developing countries and at PRESENT the lack of a portable incubator suitable for test tubes prevents the use of the multiple tube technique for FIELD testing of faecal indicator bacteria.

Table 2.2   List of equipment required for multiple tube analysis of drinking water samples

Operation	Quantity	Item	Source & Ref. No.	Cost
Sampling	30	125 ml polypropylene bottles	Gallenkamp BTK – 270 – 070T	4.95
Media preparation	1	'Speedscale' balance.	Gallenkamp BCJ – 260 – X	89.00
	10	150 ml polypropylene beakers	Gallenkamp BNH – 700 – 070J	3.60
	5	1000 ml polypropylene beakers	Gallenkamp BNH – 700 – 130R	3.65
	3	10 ml automatic tilting pipettes	Gallenkamp PMR – 781 – 080J	19.50
	3	5 ml automatic tilting pipettes	Gallenkamp PMR – 781 – 070M	18.36
	1 gross	1 oz McCartney bottles	Gallenkamp BTS – 300 – 052F	36.50
	1 gross	50 mm Durham tubes	Gallenkamp TES – 270 – 070D	3.10
	2 gross	30 mm Durham tubes	Gallenkamp TES – 370 – 030P	5.10
Sterilization	1	Pressure cooker	Prestige HiDome	25.00
	1	Electric hotplate	Gallenkamp 600S	19.50
Inoculation	20	10 ml measuring cylinders	Gallenkamp CYL – 300 – 032X	23.00
	20	10 ml beakers	Gallenkamp BNB – 300 – 030V	15.80
	20	1 ml pipettes	Gallenkamp PMF – 670 – 040B	18.50
Incubation	1 or 2*	Water bath, with lid	Grant: JB2 & LJ2	144.00 each
	3	Thermometers – 10 °C to 50 °C	Gallenkamp THL – 210 – 030C	4.50
Miscellaneous items		Bottle racks, draining racks, etc.		20.00
Media	500 g	Dehydrated MacConkey Broth		10.00
	500 g	Dehydrated Brilliant Green Broth		15.00
	500 g	Dehydrated Lactose Broth		10.00
			Total:	£489.06

* If 2 incubations at 35 °C (or 37 °C) and 44 °C are to be performed.

2.5.2 Membrane filtration technique

The membrane filtration procedure involves filtering a measured volume of sample, or an appropriate dilution of it, through a membrane filter which has a pore size of 0.45 µm and is usually made of cellulose esters. Micro-organisms are retained on the filter surface which is then incubated face upwards on a suitable selective medium containing lactose. Characteristic acid or aldehyde producing colonies develop on the membrane and these are counted as either presumptive coliform organisms or faecal coliform organisms, depending on the incubation temperature. Since gas production cannot be detected on membranes it is assumed that gas is produced by organisms that produce acid or aldehyde from lactose. The visible colonies are counted and expressed in terms of the number present in 100 ml of original sample (WHO, 1983). These colonies may be picked-off the membrane and subcultured in necessary for further identification. By incubating samples at 44 °C or 44.5 °C it is possible to determine directly the number of faecal coliforms within 24 hours, thus allowing more rapid remedial action if necessary. Since the primary objective of this chapter is to describe the detection of faecal contamination the method described below (Section 2.5.2.5) concentrates on the determination of faecal coliform organisms with incubation at 44.5 °C with currently commercially available equipment. There are however some waters which may cause difficulty with the membrane filtration technique such as:

1. The membrane soon clogs when waters with high suspended solids, turbidity or algae, are filtered. The sample could be diluted if high bacterial levels are expected, but generally the technique is not suitable for turbid waters.
2. Membranes are also unsuitable for waters containing only a few coliform organisms in the presence of many other organisms capable of growing in the media used since the latter may cover the membrane and overgrow the coliform organisms.

'Waters in which 100 ml cannot be filtered by membranes are most unlikely to be suitable for drinking purposes and, therefore, bacteriological testing of such water is generally pointless.' (B. Lloyd, personal communication, 1983)

The results given for the membrane filtration technique are not necessarily the same as those obtained by the multiple tube method although in practice they are usually comparable. McJunkin (1976) suggests 'that a skilled laboratory worker should receive a week's training and have an opportunity for further individual practice before he undertakes membrane filter analyses'. Bacteriologists do not agree on the relative skills and training needed for either the MF or MPN technique. Both techniques require training, skill and understanding before reliable results are produced.

2.5.2.1 Costs of membrane filtration

Although only limited quantities of equipment are required for membrane filtration, this is initially more expensive than the equipment used in the multiple tube method. For example, for faecal coliform testing by the membrane filtration technique the field equipment can cost at least £1865. This includes a portable field incubator able to run off a 12 Volt battery or mains supply (£1256) and a kit for field membrane filtration (£609). The consumables (membranes, petri dishes, pads and selective growth media) are also expensive. Costs can be cut down if samples are collected and incubated in a mains powered incubator. The operation of the membrane filter technique with pre-packed sterile membrane filters, sterile petri dishes with pads inside, and sterile ampoules of prepared selective growth media has become technically easier. Dr B. Lloyd is developing equipment to reduce the costs of membrane filtration, and he should be contacted for details (personal communication, 1983).

An even simpler membrane filtration process using pre-sterilized enclosed monitors cuts down possible difficulties in handling and equipment required. Unfortunately the field monitors (Figure 2.11) do not fit into the field incubator supplied by the same company although they could be used with a mains powered incubator with only the sample collection being done in the field.

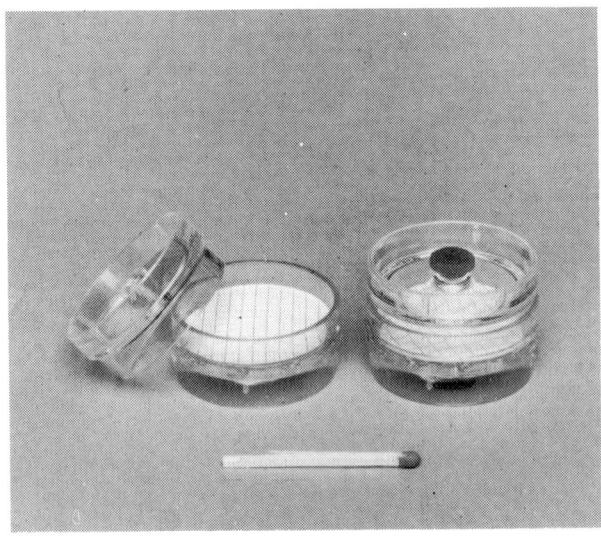

Figure 2.11 Millipore field monitors

2.5.2.2 Membrane filtration apparatus and ancillary equipment

There are several systems available for filtering, ranging from simple field monitors and syringes, through polycarbonate plastic vacuum systems with syringes, to stainless steel systems, all designed especially for the field.

Membrane filters

The membrane filters are made of cellulose acetate esters or cellulose nitrate esters and are manufactured to give a controlled pore size of 0.47 µm. The membranes measure 47 mm in diameter and have black grid marks. Several manufacturers produce them, including: Millipore, NEERI, Sartorius, Oxoid, Nuclepore and Gelman. It is worthwhile checking manufacturers' prices, especially when purchasing large quantities.

Individually sealed, pre-sterilized filters should always be purchased. They are available in various pack sizes between 100 and 1000. For example: 100 pack of HA filters from Millipore UK with absorbent pads 47 mm diameter, Cat. No. HAWG 047SO. Cost: £23.00.

Selective growth media

Growth media may be prepared from dehydrated powder or used directly from pre-measured, ready-to-use glass ampoules. Ampoules are easier to use and for field work are much more convenient. The ampoules are guaranteed for one year after manufacture when stored in a refrigerator.

Since the field method requires the determination of faecal coliforms the medium used is M – FC broth (Faecal coliforms) in 2 ml ampoules. A pack of 24 from Millipore (UK) will cost £14.00, Cat. No. M0000002F. The top of the ampoule is snapped off and medium is transferred aseptically onto an absorbent pad in a petri dish just prior to the filtration of the sample.

Absorbent pads

White absorbent pads are used to absorb the liquid media in the petri dish, and are available packed in dispenser tubes for insertion into petri dishes without even touching with sterile forceps. There are also sterile disposable plastic petri dishes available with a nutrient pad already inserted and sterilized. This cuts out another possible point of contamination, but increases the cost. Cost of 100 sterile petri pads in dispenser tube PD 1004750 £14.00, from Millipore.

Petri dishes

Sterile plastic petri dishes are recommended for use with the field incubator, as noted above they are also available with absorbent pads already in them. They have a diameter of 50 mm. Cat. No. PD 100004700 Sterile Petri Dishes per 100 Cost £11.00.

Cat. No. PD 1004750 Sterile Petri Dishes and pads per 100 Cost £14.00.

Mara (1974) describes the process of re-sterilizing the plastic disposable petri dishes, namely removing the filter and absorbent pad with forceps, soaking the dish in 5% V/V Formalin for 1 hour, rinsing thoroughly in tap water, and soaking in absolute alcohol overnight. The alcohol is removed by evaporation.

Pipettes
Pre-sterilized 1 ml and 10 ml graduated plastic pipettes are available and are useful for dilution purposes. They are available in packets of 25. Cat. No. XX6300135 Sterile pipettes 1 ml, cost £10.00.

Sterile phosphate buffer solution
This is used to dilute samples and to rinse the membrane filters. The solution must be sterilized before use.

Stock Solution I: 34.0 g of potassium dihydrogen phosphate (KH_2PO_4) dissolved in 500 ml of distilled or deionised water. Adjust the pH to 7.2 with 1N sodium hydroxide (NaOH). Then dilute to 1 litre to produce a stock solution. This is best stored in a refrigerator and should be discarded if it becomes turbid.

Stock Solution II: 60 g of magnesium sulphate ($Mg\ SO_4.7H_2O$) dissolved in 1 litre of distilled or deionised water.

Working phosphate buffer solution (for dilution bottles and filter rinsing purposes). Add 1.25 ml of stock solution I and 5.0 ml of stock solution II to a bottle of distilled or deionised water and dilute to 1 litre. This must be sterilized before use either in the pressure cooker (best in four batches of 250 ml so that you have a spare bottle if one is contaminated) or by filtering through a Millipore GS (0.22 m pore size) or Millex membrane filter, especially useful for filter rinsing purposes in the field.

Portable MF-Millipore petri dish incubator
The aluminium block MF-Millipore incubator (Figure 2.12) is a non-water bath incubator and is the only one suitable for faecal coliform analysis, achieving the desired 44.5 °C ± 0.2 °C either in the field or on the bench. It holds up to 30 petri dishes and can be run off 12 volts DC (car or truck batteries), and 110 volts or 220 volts AC (depending on mains supply in the country). For field use adaptors to plug into cigarette lighter sockets in the car are available and it is useful when doing field incubation to have a spare battery available to supplement the vehicle battery. The unit draws 15 watts maximum and some car batteries may not be able to sustain a 12 hour discharge for long at

24

1.25 amps. This can be alleviated by plugging the unit overnight into a mains supply (if it is reliable and not subject to failures). This incubator has been used successfully for field work in Botswana (Hutton and Lewis, 1980).

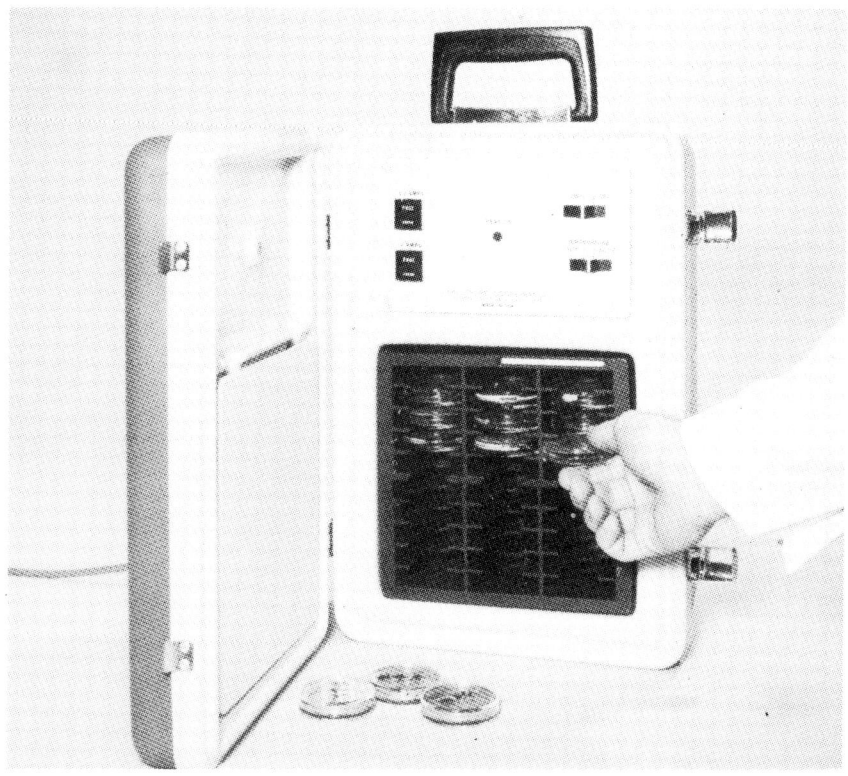

Figure 2.12   Aluminium block MF-Millipore incubator

It can operate at 44.5 °C or 35 °C for Total Coliform counts if necessary. XX6300400 12 volt DC/110 AC and XX6300405 12 volt DC/220 VAC XXFE00100: Cost £1256.00.

Faecal coliform field kit (Figure 2.13)
This kit comes in a hard wearing case and contains:
stainless steel filter assembly;
stainless steel sample cup;
syringe with a 2-way valve;

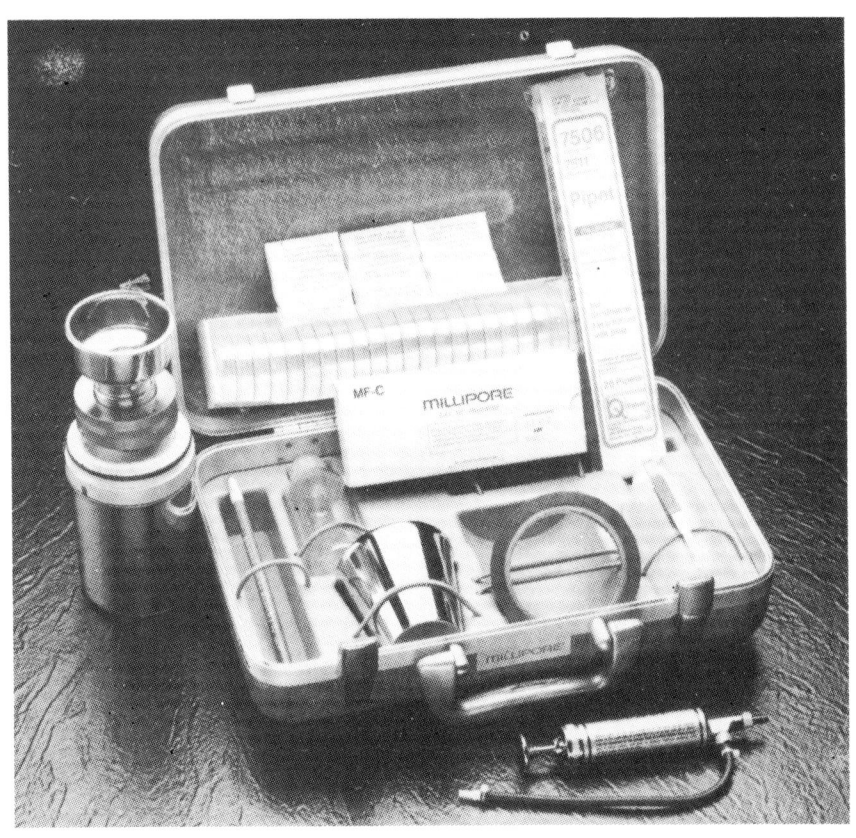

Figure 2.13 Millipore faecal coliform field kit

smooth tipped stainless filter forceps;
plastic alcohol bottle;
plastic tubes with adaptors;
100 sterile plastic petri dishes;
sterile plastic pipettes;
24 × 2 ml ampoules of M-FC broth;
100 Type HAWG sterile 0.47 μm membrane filters with absorbent
pads 47 mm diameter;
1 red wax pencil (for labelling petri dishes);
1 roll sealing tape;
Cost: £609.00.

This kit plus the incubator will enable field tests for faecal coliform
bacteria to be carried out on some 100 samples.

26

It is suggested that at least two more pairs of forceps, another syringe and a sample cup be also purchased if extensive field testing is to be undertaken.

Sterile rinse water

If sufficient buffer rinse is not available or is used up, sterile rinse water can be prepared using a 0.22 μm filter, and a syringe (see Figures 2.14 – 2.17). The rinse water is purified by passing through the Millex 0.22 μm filter which removes all particles and dead or alive bacteria.

Millex Filter Unit SLHA 025 BS, Cost £42.00 per 50.
50 ml Syringe Cat. No. XX1105005, Cost £14.00 per 5.

Figure 2.14   Draw into the syringe 50 cc (50 ml) of water from any natural source

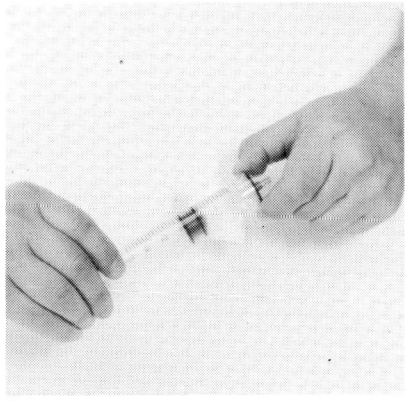

Figure 2.15   Attach Millex filter unit aseptically

27

Figure 2.16   Filter unit ready for use to produce sterile water

Figure 2.17   Force water through the sterilizing filter onto the inner walls of the filter holder funnel

2.5.2.3   Volume of water to be examined and dilution procedures

When counts of bacteria are expected to be high the sample must be diluted to obtain a count on the incubated membrane filter of between 20 and 120 visible colonies. The dilutions are made using sterile pipettes capable of measuring 1 ml accurately to sterile buffer solutions in 15 × 150 mm clean screw cap culture tubes for 1:10 dilution, and 100 ml screw cap dilution bottles for 1:100 dilutions (Figure 2.18).

28

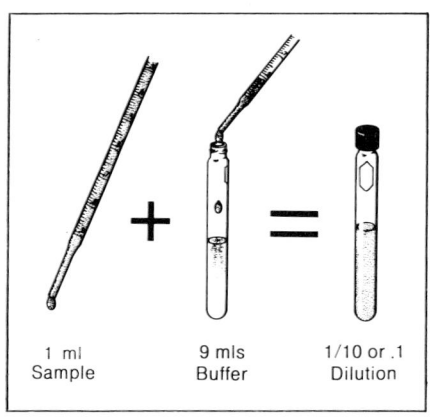

| 1 ml | 9 mls | 1/10 or .1 |
| Sample | Buffer | Dilution |

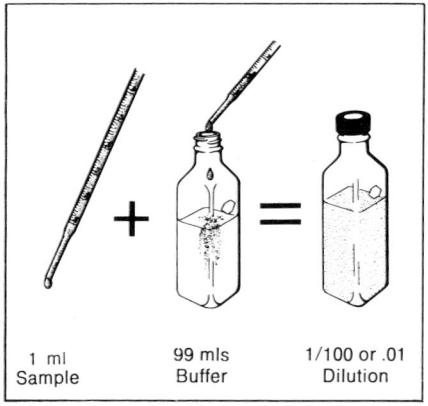

| 1 ml | 99 mls | 1/100 or .01 |
| Sample | Buffer | Dilution |

Figure 2.18   Dilutions 1/10 and 1/100

To prepare dilution blanks is simple. It is best to sterilize the buffer and its container together in the pressure cooker. Using a measuring cylinder add 102 ml of phosphate buffer solution (see below) to the screw cap dilution bottles and for the culture tubes 9.5 ml. Place them in the pressure cooker for 15 – 20 minutes for sterilization with the tops LOOSELY screwed on (or they explode!). The excess buffer will evaporate and after they have cooled close the tops tightly. Store the sterile blanks in a cool dark place to prevent photodegradation.

Dilutions must be made with sterile pipettes each time and aseptically. A review of dilutions can be made if bacteria levels are expected to be high. The sample sizes below may be used as a rough guide:

Table 2.3. Sample size for various waters

Treated drinking water	50 ml		100 ml	
Well and borehole water	10,	50,	100 ml	
Unpolluted surface waters	1,	5,	10,	50 ml
Polluted surface waters	0.05,	0.1	0.5,	1.0 ml

When small volumes are taken from dilution bottles and tubes the filtered sample must be rinsed on the membrane filter with at least 30 ml of sterile buffer to even out the distribution of faecal indicator organisms on the membrane filter surface.

2.5.2.4 Equipment check list before leaving for field work
1. Sterile sample bottles containing sodium thiosulphate, and sampler if necessary.
2. Sterile sample dilution bottles and tubes.
3. Filter holder sterilized either by:
    (a) autoclaving wrapped in Kraft paper
    (b) boiling water for 3 − 5 minutes
    (c) immersing in absolute alcohol and drying
    (d) using methanol on wick and replacing cover over unit.

Formaldehyde formed by the incomplete combustion of methanol sterilizes the holder, see diagrams below:

Figure 2.19   Use the tips of Millipore forceps to get a start on opening the stainless steel filter holder

Figure 2.20   A ½-cap of methanol is poured on the asbestos ring as the entire unit is slowly revolved

Figure 2.21 Don't wait – ignite the methanol immediately

Figure 2.22 As soon as the wick is ignited, replace the receiver flask over the holder flask and wait 15 minutes

4. Sufficient sterile membrane filters, pads and petri dishes.
5. Faecal coliform field kit and spare syringe, forceps and sample cup.
6. Sterile media – MFC broth.
7. Gas torch, alcohol bottle and matches.
8. Incubator. It is useful to run it up to 44.5 °C from a mains supply before plugging into the vehicle supply.
9. Sterile buffer and rinse solution.
10. Millex equipped syringe and spare Millex holders.
11. Wipex cloths and tap cleaning materials.
12. Record book for sample times, date and location, and wax pencil or laboratory marker pen.

2.5.2.5 Method: Faecal coliform at 44.5 °C, using faecal coliform test kit

Equipment needed
Millipore faecal coliform test kit XXFC00100 and equipment, portable incubator, XX6300405, gas torch (for sterilizing taps). Wipex cloths (for cleaning work area and hands).

Procedure
1. Open a sterile petri dish using the forceps previously sterilized by keeping them in at least 3 cm of ethyl or methyl alcohol and flaming them in a flame and allowing them to cool. Do not hold the forceps in the flame longer than necessary to ignite the alcohol. Invert the top and

31

place it alongside the base. Place a sterile absorbent pad into the lower half of the petri dish using either the forceps or the sterile dispenser. Replace the forceps in the alcohol bottle.

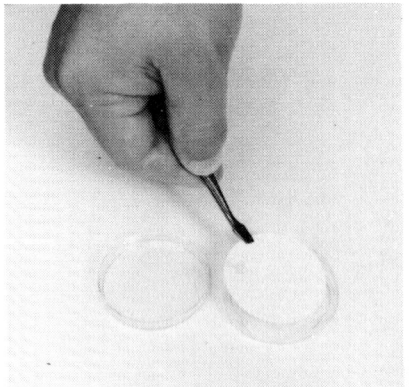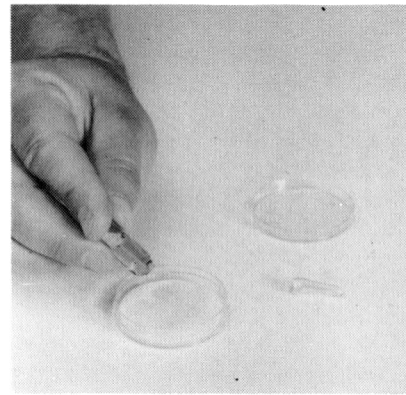

Figure 2.23

2. Break off the top of a 2 ml glass ampoule of MFC broth and tap out or shake entire contents onto the absorbent pad. Do not allow the ampoule to touch the petri dish or absorbent pad. Replace the cover of the petri dish.

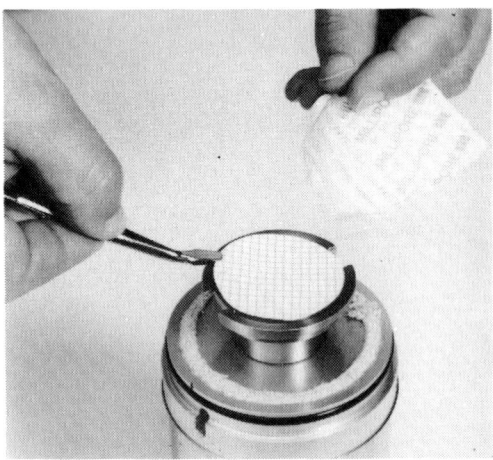

Figure 2.24

3.   Open the sterile filter holder assembly and, touching only the outside of the holder, remove the funnel from the top and place on a clean area. Using sterile forceps remove a sterile type HA membrane filter from its packet by the edge and place it grid side up centrally on the filter support screen. Replace funnel assembly tightly taking care not to damage the membrane.

4.   Shake the water sample (if in a bottle) vigorously up and down at least 25 times.

5.   If the sample is 10 ml or more pour the sample into the funnel, wetting the filter.

For smaller sample sizes 20 – 30 ml of sterile buffer should be poured in first to give an even distribution of sample above the filter

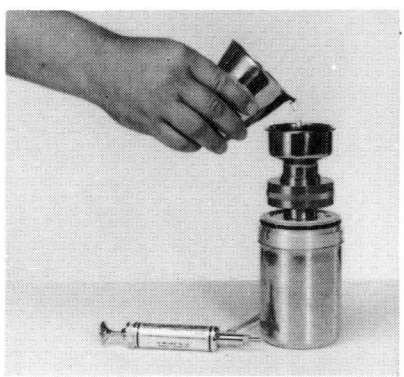

Figure 2.25

6.   Apply suction by pumping with the vacuum syringe and suck the sample through the filter.

 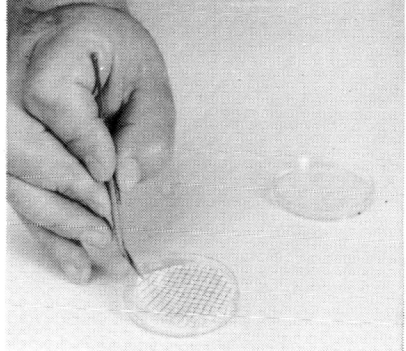

Figure 2.26

33

7. Rinse the funnel with 20 – 30 ml of sterile buffer. Suck the rinse through the filter. Repeat the rinse step with another 20 – 30 ml rinse buffer.

8. Disassemble the filtration assembly. Use sterile forceps to remove the membrane filter from the base of the filter holder. Place the filter, grid side up onto the nutrient medium in the petri dish (prepared in step 2) with a slight rolling action so that air is not trapped underneath it. If air bubbles are seen reposition the membrane filter on the pad. Place the cover on the petri dish tightly. Label the dish with date, time and sample reference. (See note on Field Operation.)

 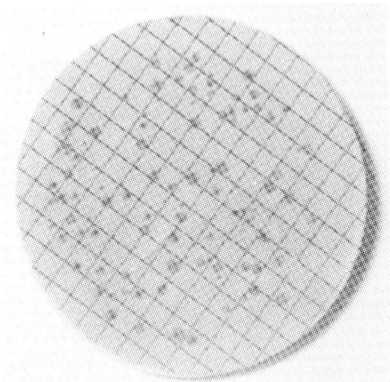

Figure 2.27

9. Place the closed petri dish upside down in the incubator and incubate at 44.5 °C for 24 hours.

10. After incubation remove the cultures and count the number of colonies having a blue colour as the count of faecal coliform bacteria in the volume of sample taken and then calculate the number per 100 ml.

Field operation
After completion of each filtration, it is possible to proceed directly to further filtrations without re-sterilizing, *BUT* if more than 15 minutes elapse between the filtration of successive samples the unit must be re-sterilized as described earlier in the equipment checklist.

2.6 Dipslide tests for bacteria
These are pre-sterilized disposable kits which contain growth media and are used to measure large quantities of bacteria (greater than 100 colonies (100 ml)) in a small quantity of water (1 ml or 5 ml).

34

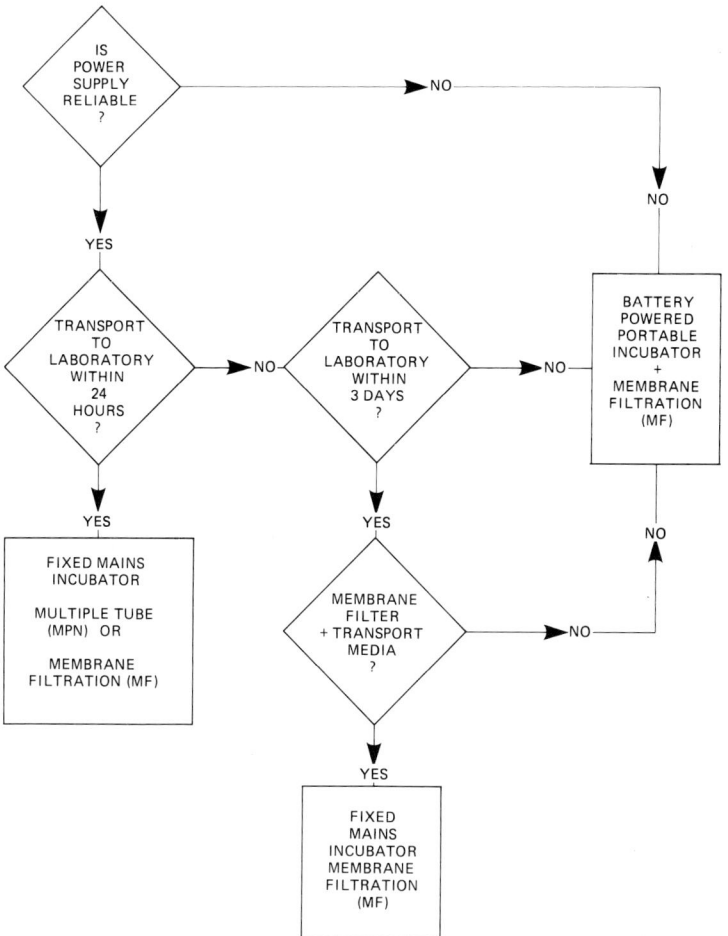

Figure 2.28   Algorithm for choice of procedures for bacteriological testing of water in developing countries

Oxoid manufacture a dipslide coated with a non-selective nutrient agar on one side and MacConkey agar on the other. The dipslide is removed from its container and dipped into the sample (sewage or river) for 5 seconds and then replaced into its container. The container is then incubated at 37 °C and after 24 hours the counts of colonies on the MacConkey agar estimated using standard charts supplied by the manufacturer. Rough counts of coliform organisms in the range 10 000 to 10^9 per 100 ml can be made. Cost per 20 dipslides £4.63.

Millipore also produce a dipslide counter using an ingenious membrane filtration procedure which filters 1 ml of sample through a membrane by capillary action of an absorbent pad (which also contains dehydrated nutrient media). The pad is replaced into its container and incubated at 44.5 °C for faecal coliform organisms for 18 – 24 hours. Blue or blue green colonies are counted and reported in the range of 100 – 50 000 colonies per 100 ml. Cost per 20 dipslides (MC0000020) £50.00.

These dipslides may be useful for monitoring sewage effluents or gross river pollution, but are neither accurate nor sensitive enough for monitoring drinking waters.

They still require the use of an incubator although some workers tape the slides to their bodies for incubation at 35 °C.

2.7 Summary

The choice of techniques and apparatus for bacteriological testing of faecal contamination is wide and the costs of testing are similarly very variable. The cost of basic multiple tube method apparatus is obviously much lower than the cost of portable incubators and membrane filtration apparatus. An estimate based on the list of equipment for multiple tube analysis of drinking water supplies (see Table 2.2) compiled by Mara in Cairncross and Feacham (1978) with 1983 prices totals £489.00.

The choice of procedure is governed by the local situation but the algorithm, Figure 2.28, summarises some of the factors (apart from economic constraints) which must be considered.

It is once again important to emphasise that the bacteriological quality of drinking water as assessed by its freedom from contamination by faecal material is THE MOST RELEVANT WATER QUALITY GUIDELINE to be considered in developing countries.

3. Chemical Analysis

The main advantages of chemical analysis in the detection of water pollution are its rapidity compared with bacteriological analysis and pollution not caused by or reflected by organisms.

Field determinations cannot be expected to be as accurate as those performed under good laboratory conditions. The limits of detection and reproducibility will not be comparable in many determinations (although they often are) but these factors are often overlooked in favour of rapid, on site determination of possible pollution.

Many parameters of water quality change on transportation (and even in sampling) to a laboratory, and field tests can overcome the need for preservation of samples prior to laboratory analysis.

There are limitations to the application of field chemical analysis especially at low levels of determinand. The limit of detection of some field methods is higher than the levels of interest (usually the WHO Guideline) especially in the cases of heavy metals such as lead, chromium, arsenic and even manganese. The methods for these elements are included to allow field tests to be made (to detect gross pollution) but suitable warnings are given in the relevant sections. In such cases there is an obvious need for more research to be directed at improving reagents and methodology. Where there is uncertainty in such determinations, samples should be taken (and preserved if necessary, see Section 3.1) for subsequent analysis as soon as possible in a reliable laboratory.

There are some determinations which do not lend themselves to field techniques, such as those for sodium and potassium. These require a flame photometer, power, gas and air supplies which, although they may be fitted into a mobile laboratory, are not portable. Further research may develop field flame photometry but at present we are unable to determine an ionic balance of major constituents in the field. Samples should be taken in plastic bottles for subsequent laboratory analysis.

Techniques
The main techniques of field chemical analysis for water are:

(1)   Colorimetric   Paper test strips
                            Printed colour comparator cards
                            Coloured cubes
                            Discs and comparators
                            Field spectrophotometers

(2)	Titrimetric	Tablets
		Dropping burettes
		Digital titrators
(3)	Specific Ion	for fluoride, dissolved oxygen, pH
	Electrometric	nitrates.

Interferences

Many analytical procedures are subject to interference from other substances present in the sample (AWWA, 1980). Interferences can give either higher or lower results and can even be compensatory. It is IMPERATIVE to note any abnormal colours or suspicious reactions and refer closely to the method to see if the detailed instructions supplied with most kits mention possible interferences.

In several cases the kit reagents are formulated to complex, oxidise or reduce interfering ions or compounds but these interfering agents may be present in larger amounts than the reagents are designed to deal with.

Sometimes pretreatment procedures are described in the test kits and these may be sufficient to overcome the problems, but be on your guard for interference especially with water of unknown composition. Dilution of the sample commonly dilutes the interfering agent, but it may also cause problems in detecting the substance being determined. Turbid and coloured samples cause problems especially in colorimetric procedures. Filtration or settlement may alleviate these difficulties, but may introduce adsorption or contamination problems especially with lead and other trace metals. Free chlorine also causes problems because it bleaches and destroys indicators and oxidises other reagents used in field techniques. It is usually removed by pretreating the sample with sulphite, thiosulphate or arsenite solutions.

Some methods, particularly those for trace metals, lead and arsenic may not recover all forms of the determinand present in the sample without some form of pretreatment. In some cases the pretreatment may not be practical under field conditions, and these limitations must be noted. Many organically bound metals may need boiling or taking down to fumes with nitric and hydrochloric acid and this will not be practical in most cases.

Procedures

Considering the points raised about interference, accuracy and limits of detection, it is important to emphasise that the details of procedure be STRICTLY followed. It is important to add the reagents in the sequence suggested in the test kit procedure, to shake for the specific period, and to wait the prescribed time for colour development to occur. The

38

procedures have been well tested and tried by all the manufacturers mentioned in this publication and there are good analytical reasons for these methods to be followed closely.

If the instructions are lost most manufacturers will be happy to replace them free of charge. The author recommends that all instructions sheets are copied and that a spare copy be stored in a safe place should the field copy be lost.

Quality control
Many of the field kits are supplied with pre-tested reagents which meet stringent quality checks by the manufacturers. With replacement reagents it is recommended that tests on a blank (distilled or deionised water) and on a standard solution be made for every new batch of reagents, and especially in cases where the reagents may be old. It is useful to do these calibration checks at suitable opportunities when the user has access to a reliable laboratory. Some manufacturers market standard solutions for calibration purposes.

3.1 Sampling and preservation techniques
Most of the test kits and methods described in this chapter are designed for on site analysis of water. Often the reaction vessels themselves are the sampling containers. As noted in the bacteriological section, the samples taken must be representative of the water body under investigation.

It is important not to contaminate the sample by contact with other materials in the sampling or transfer process. For example, lead may be lost rapidly from unacidified samples by adsorption on container walls, or contamination by metallic sample collectors may add trace metals to the sample.

There will be cases where analysis cannot be carried out immediately and samples may need more accurate confirmatory tests in a laboratory.

There are several procedures for preserving samples prior to laboratory analysis – generally they involve slowing down biochemical reaction rates by refrigeration, adding chloroform or mercuric chloride, or adjusting the pH with acids or alkali. Table 3.1 gives some details of preservation techniques. For cases where the data is not clear on preservation, or where the sampler is in doubt he should contact a qualified laboratory analyst at the receiving laboratory, or refer to the test kit instructions. It is important to carry out field determinations for nutrients such as the nitrogen species, phosphates etc. because their concentrations change on storage. The determination of gases such as dissolved oxygen, carbon dioxide, and chlorine must be carried out immediately on sampling to have any value.

Table 3.1 Preservation techniques

Determination	Container	Preservation	Max. Storage
Acidity	P, G	Refrigeration	14 days
Alkalinity	P, G	Refrigeration	14 days
Ammonia	P, G	Add H_2SO_4 to pH $<$ 2 refrigerate	28 days
Arsenic	P, G	ANALYSE IMMEDIATELY	
Boron	P	None required	28 days
Calcium	P, G	Add HNO_3 to pH $<$ 2	6 months
Carbon Dioxide	P, G	ANALYSE IMMEDIATELY	—
Chloride	P, G	None required	6 months
Chlorine	P, G	ANALYSE IMMEDIATELY	
Chromium	P, G	ANALYSE IMMEDIATELY	
Conductivity	P, G	ANALYSE IMMEDIATELY	
Copper	P, G	ANALYSE IMMEDIATELY 0.5 ml 50% Hcl/100 ml and ANALYSE IMMEDIATELY	
Cyanide	P, G	Add NaOH to pH $<$ 12, Refrigerate in dark	14 days
Fluoride	P	None required	28 days
Iron	P, G	ANALYSE IMMEDIATELY or 1 mL Conc HCl per 100 ml sample	14 days
Lead	P, G	Acidify conc HNO_3 to pH $<$ 2	6 months
Magnesium	P, G	Add HNO_3 to pH $<$ 2	6 months
Manganese	P, G	ANALYSE IMMEDIATELY or Add HNO_3 to pH $<$ 2	2 days
Nitrate	P, G	Add H_2SO_4 to pH $<$ 2, refrigerate	2 days
Nitrite	P, G	ANALYSE IMMEDIATELY or freeze at $-20°C$	2 days
Odour	G	ANALYSE IMMEDIATELY	6 hours
Oxygen	G	ANALYSE IMMEDIATELY	
pH	P, G	ANALYSE IMMEDIATELY	
Phosphate	G	ANALYSE IMMEDIATELY or refrigerate	2 days
Redox Potential, Eh	IN-SITU P	ANALYSE IMMEDIATELY or freeze	
Sulphate	P, G	Refrigerate	28 days
Taste	G	ANALYSE IMMEDIATELY	
Temperature	P, G	ANALYSE IMMEDIATELY	
Turbidity	P, G	ANALYSE SAME DAY, STORE IN DARK	2 days

Based on APHA (1980).   (P = Plastic G = Glass Borosilicate).

In most cases plastic bottles with screw caps are suitable containers for sampling and transport. It is useful to have some sample bottles in different sizes to allow for the different preservation techniques. About $2 - 2½$ litres is sufficient for major ion analysis in a laboratory, with smaller samples taken for special analyses. The amounts needed depend on the techniques being used in the laboratory.

Samples should be securely fastened and labelled with time, date, location, name of sampler, preservation technique used, type of

40

analysis required. It is useful to equip and supply the sampler with waterproof marking pens and masking tape for sticking onto bottles. Tie-on labels always detach themselves.

Bottles may be cleaned with detergents and/or concentrated hydrochloric acid rinsed with tap water and finally with distilled or deionised water. Many problems of sampling are due to inadequate cleaning of glassware. The cleaning agents used should not contain the determinand, for example do not use detergent for cleaning if the sample is being analysed for phosphate; similarly, for nitrates do not use nitric acid, for chlorides do not use hydrochloric acid. The bottles must be physically clean, and bottle brushes are useful in removing stubborn deposits. Do not use a dirty bottle to sample the water. The sample bottle should be rinsed 3 times with sample before the bottle is filled and top secured. Always carry some spare tops for bottles.

Bottles should be shipped or transported in cases with compartments for each bottle. They should be protected from strong sunlight and vibration.

All the analyses described in this chapter are best carried out on site to have their greatest value.

3.2 Calcium (and hardness)

In older texts greater emphasis was placed on 'hardness' of waters. 'Hardness' was the term used to define the soap-destroying properties of water and the formation of objectionable curd-like precipitates of calcium and magnesium stearates and oleates. The determination of hardness was carried out using standard soap solutions and observing when a permanent lather was obtained.

A more accurate procedure for the determination of hardness was developed using EDTA (Ethylene Diamino Tetra Acetic Acid) and this reagent allows the differential estimation of calcium and magnesium. The 1980 (15th) Edition of APHA/AWWA/WPCF Standard Methods suggests that 'the preferred method for determining hardness is to compute it from the results of separate determinations of calcium and magnesium'.

HARDNESS = CALCIUM + MAGNESIUM

This is valid for most waters although in terms of the original definition of hardness some other metals can also contribute to hardness if they occur in significant amounts. These are iron, aluminium, manganese, strontium, and zinc. However, for most practical purposes their contribution to hardness can be neglected.

Hardness was often expressed in terms of mg/l as calcium carbonate because calcium carbonate was a familiar common chemical, was available as a useful primary standard, and had a convenient mole weight of 100. This has led to a great deal of confusion since degrees of hardness were defined by several countries. The conversion factors below may assist in interpreting older data:

1 $^{\circ}$dH	1.25 $^{\circ}$eH	1.79 $^{\circ}$fH	=	17.9 mg/l $CaCO_3$
German degree of hardness	English degree of hardness	French degree of hardness		

and since

$$100 \text{ mg/l } CaCO_3 = 40 \text{ mg/l } Ca = 24.3 \text{ mg/l } Mg^{++}$$
$$17.9 \text{ mg/l } CaCO_3 = 7.16 \text{ mg/l } Ca^{++} = 4.35 \text{ mg/l } Mg^{++}$$

The Ministry of Health on Water Softening in the UK (1949) gave an arbitrary classification of water as follows:

Description of Water	Hardness, mg/l $CaCO_3$
Soft	0 – 50
Moderately soft	50 – 100
Slightly hard	100 – 150
Moderately hard	150 – 200
Hard	200 – 300
Very hard	over 300

It should be apparent that a great deal of confusion can occur when we talk of hardness of water.

Since calcium and magnesium salts are present in water as ions it is far simpler to report their concentrations as calcium and magnesium ions according to APHA (1980).

Calcium in natural waters is due to the passage of rainwater through or over deposits of calcium-rich rocks such as limestone, dolomite, and gypsum, or cementing materials in other rocks. The range of calcium concentrations is wide, from zero to several hundreds of milligrams per litre depending on the water source and its treatment. Low levels of calcium may be of benefit in a water by laying down a protective layer of calcium carbonate on pipes and tanks. Large amounts of calcium salts can precipitate on heating to form scale deposits in boilers, pipes, and cooking pots with problems of blockages and loss of thermal efficiency.

It may be necessary to 'soften' or remove calcium salts from water by the lime-soda process, reverse osmosis or ion exchange.

Since calcium carbonate may precipitate from water samples, field tests are particularly useful for calcium determinations.

The field methods use EDTA (in solution or tablet form) and various indicators (such as 'Calcon' Eriochrome Black T, HSN Patton and

Reeder's reagent or Murexide in solution or powder or tablet form) and a high pH buffer (in powder or tablet form).

The titration equipment in laboratory use is difficult to adapt to field work because glassware and especially burettes and pipettes are easily broken. Titration can be carried out by counting drops of reagents from rubber bulb droppers, using reagents in tablet form such as Palintest tablets. More recently HACH have developed a digital titrator. The HACH digital titrator is an interesting development and consists of a precision dispensing device using concentrated standardized solutions in compact cartridges. The main features are illustrated below.

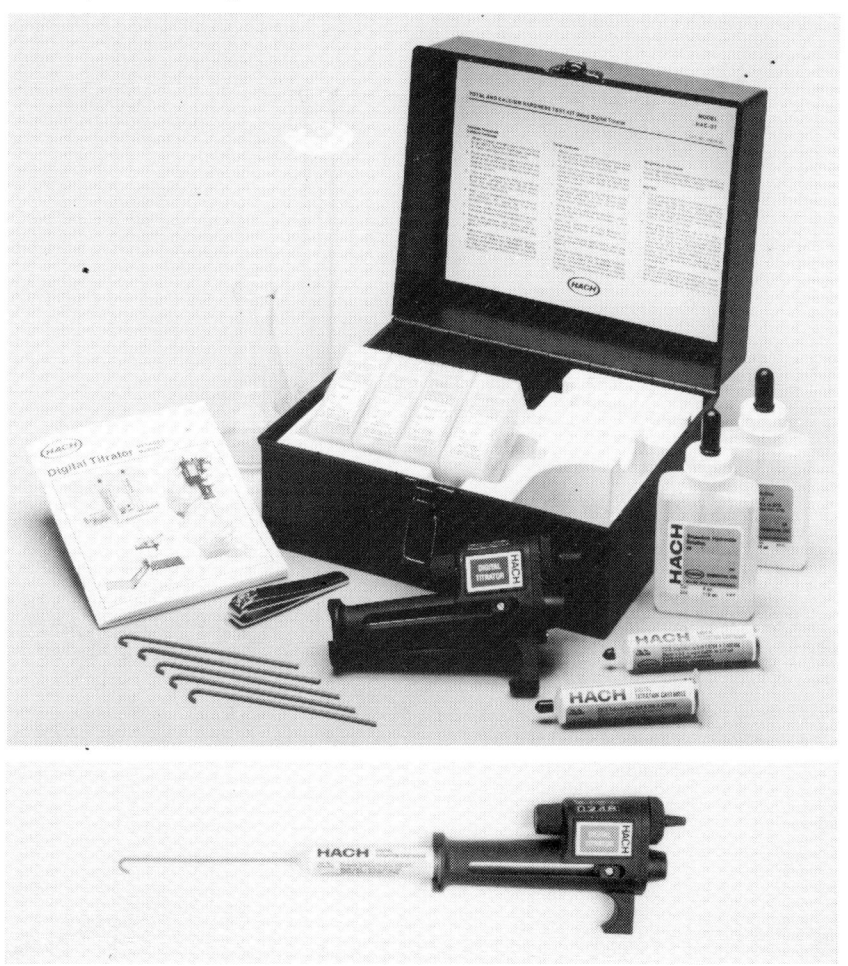

Figure 3.1 The HACH digital titrator

43

The specification below is taken from the HACH Digital Titrator Instruction Manual (1977).

'A main drive screw in the Digital Titrator controls a plunger which forces the solution from a titration cartridge in a carefully regulated flow. The titrator body is constructed of precision-moulded, heavy-duty, chemical and impact-resistant acetal plastic. Accuracy is rated at ±1% for a multiple turn titration.

Titration solutions are packaged in disposable polypropylene plastic containers with Teflon covered neoprene seals and vinyl resealable closures to cover the cartridge tips. Each cartridge contains approximately 13 ml of titrating solution, sufficient for 50 – 100 average titrations. Titrating solutions (see list on pages 10 – 11) are controlled to ±0.5% normalities with normality and tolerances listed on label. Solution concentrations are designed for titrations of 10 – 35 turns of the large knob, with the digits appearing in the counter window corresponding to the sample concentration in mg/l.

Both portable and fixed-position titrations are possible with the Digital Titrator. The instrument has a grip for hand-held operation as well as a removable shaft that clamps to a laboratory stand for stationary set ups.

Packaged in a black plastic case, each Digital Titrator comes with 3 delivery tubes (one bent and two straight stem), instruction manual and an aluminium mounting rod.'

The number of cartridges available is increasing so many titration procedures may be performed in the field.

The initial cost may be high but it can be used for several applications and its robust construction will give it a long working life.

Interferences
Orthophosphate precipitates calcium at the pH of the test. Strontium and Barium interfere with the calcium determination and alkalinity in excess of 300 mg $CaCO_3$/l may cause an indistinct end point in hard waters. Iron also interferes and the use of CDTA may be necessary. Other divalent metals may also interfere.

3.2.1 Tablet method
Supplier: Wilkinson and Simpson Ltd.
Kit description: Palintest Calcium Test Kits.
Range: 0 – 500 mg/l $CaCO_3$.

Sample size: 200 ml, 100 ml, 50 ml.
CS119 Compact Kit 20 tests.
SS219 Standard Kit 100 tests.
Uses tablet reagents, shaking bottle and measuring cylinders.
Tablets available in bottles or foil wrapped.
Cost: CS119 £6.75
    SS219 £21.25

3.2.2 HACH digital titrator
Supplier: Camlab Ltd.
Kit description: HACH Calcium and Total Hardness Test Kit.
High and low range with Digital Titrator. Model HAC − DT.
Cat. No. 20639 − 00.
Ranges: 0-100 mg/l and 0-1000 mg/l.
Calcium and Hardness as $CaCO_3$ = 0-40 mg/l and 0-400 mg/l as
Calcium Ca^{++}.
Sample size: 100 ml.
Titration using EDTA cartridges (or CDTA if iron interferes).
No. of tests per kit: 200.
In black plastic carrying case with all equipment required and instruction booklet and digital titrator.
Cost: £146.00.

3.3 Magnesium
Magnesium is a common constituent of natural waters; its concentration ranges from zero to several hundred milligrams per litre. The source of magnesium in water is the chemical weathering of rocks such as dolomite and also silicate minerals found in igneous rocks. It is usually found in lower concentrations than calcium due to the greater abundance of calcium in the earth's crust. It is a major contributor to 'hardness' and like calcium forms scales and deposits on heating and may have to be removed by softening processes.

Concentrations of magnesium above 150 mg/l, especially if present with sulphate, can cause gastro-intestinal irritation and diarrhoea.

The field tests for magnesium are usually based on the difference between Total Hardness and calcium concentration as determined by EDTA and suitable indicators.

Interferences
Iron does not interfere up to 15 mg/l. Above this level it causes a red-orange to green end point that is sharp and usable up to 30 mg/l iron. Manganese titrates directly up to 20 mg/l but masks the end point above this level. Adding a 0.10-g scoop of hydroxylamine hydrochloride monohydrate raises this level to 200 mg/l manganese.

Copper and aluminium interfere at levels of 0.10 and 0.20 mg/l, respectively. Cobalt and nickel interfere at all levels and must be absent or masked. A 0.5-g scoop of potassium cyanide removes interference from up to 100 mg/l copper, 100 mg/l zinc, 100 mg/l cobalt and 100 mg/l nickel. A 1.0-g scoop raised the permissible aluminium level to 1 mg/l. Excess potassium cyanide has no effect on test results.

Barium, strontium and zinc titrate directly. Orthophosphate causes a slow end point, and polyphosphate must be absent for accurate results to be obtained. Acidity and alkalinity at 10 000 mg/l (as $CaCO_3$) do not interfere. Saturated sodium chloride solutions do not give a distinct end point.

3.3.1 Tablet method
Supplier: Wilkinson and Simpson Ltd.
Kit descriptions: Palintest Hardness Test Kits.
Several kits covering several ranges and numbers of tests.
CS118 Hardness VLR Compact Kit 0-5 mg/l as $CaCO_3$, 20 tests.
Cost: £6.75.
SS218 Hardness VLR Standard Kit 0-5 mg/l as $CaCO_3$, 100 tests.
Cost: £21.25.
CS117 Hardness BW(LR) Compact Kit 0-50 mg/l as $CaCO_3$, 20 tests.
Cost: £6.75.
SS217 Hardness BW(LR) Standard Kit 0-50 mg/l as $CaCO_3$, 100 tests.
Cost: £21.25.
CS116 Hardness Compact Kit 0-500 mg/l as $CaCO_3$, 20 tests.
Cost: £6.75.
SS216 Hardness Standard Kit 0-500 mg/l as $CaCO_3$, 100 tests.
Cost: £21.25.
WS438 Hardness (Eriochrome) Comparator Test Kit 0-60 mg/l as $CaCO_3$, 200 tests, using Lovibond 2000 Comparator and Disc 4/38.
Cost: £52.65.

In order to determine magnesium concentrations both Total Hardness and calcium hardness must be determined (see section on Calcium for details of Ca^{++} testing).

3.3.2 The HACH digital titrator
Supplier: Camlab Ltd.
Kit description: HACH Calcium and Total Hardness Kit HAC-DT as described in Calcium section.
Cost: £146.00.

3.4 Iron
Since iron is present in nearly all rocks, sands and soils almost all natural waters contain some dissolved iron often only in traces, but

sometimes up to 50 mg/l. Iron in its dissolved form exists as ferrous ions, Fe^{++}. On exposure to air or oxygen the ferrous ions (Fe^{++}) are oxidized to ferric ions (Fe^{+++}) which react with hydroxyl ions in the water to form insoluble or colloidal brown ferric hydroxide. Many groundwaters when freshly pumped to the surface are clear and bright, but if iron is present on exposure to air the water becomes opalescent, discoloured with a brown precipitate.

This conversion of ferrous (soluble) iron to ferric (insoluble) iron may occur on transporting the sample to the laboratory so field tests for iron are valuable.

The problems created by high iron concentrations are mainly aesthetic but can seem very serious to the consumer. Rice and vegetables develop brown deposits when cooked in water containing iron. Tea develops an inky colour and the water tastes bitter or metallic. Clothes develop rust stains when washed in water containing iron. Porcelain fittings become discoloured.

For the water engineer ferric hydroxide deposits in dead ends and in pipes may lead to blockages of mains, meters etc. Bacteria may multiply in the deposits allowing the introduction of iron sulphides and hydrogen sulphide into the supply, leading to complaints about odour, colour and taste. The WHO guide level is 0.3 mg/l Fe on organoleptic grounds.

Some industries, such as papermaking, food and beverage manufacturing, dyeing and ice making require iron free water (less than 0.2 mg/l).

The field methods of determining iron depend on the development of colours produced by the reaction of iron with orthophenanthroline or thiogycollic acid. Orthophenanthroline is preferred in most cases, but there are new developments underway from Wilkinson and Simpson who are introducing a tablet/colorimetric test incorporating thiogycollic acid and also PPST reagent (Wilkinson and Simpson, personal communication, 1983).

Orthophenanthroline reagent only reacts with ferrous iron ions and most reagents include an acidification and reduction component to reduce ferric ions to ferrous ions prior to analysis. The total iron content can be determined in most cases by adjusting the pH to less than 3 using hydrochloric acid, but difficulties may be encountered if large amounts of ferric precipitates are present. Organically complexed iron is not determined directly.

Interferences
Oxidising agents such as chlorine interfere with these procedures and should be removed with thiosulphate. Amounts of copper and cyanide greater than 1 mg/l may interfere.

3.4.1 Visual zero cost method (IDRC, 1981)

Equipment required:
(a)   clean white porcelain vessel (cup), or
(b)   clean white paper.

Procedure:
(a)   Pour fresh sample into cup and observe over a period of time if a reddish yellow or brown deposit or precipitate develops. If so, iron is present.
(b)   Drip a few drops of sample onto white paper and allow them to dry. Check the edge of the watermark for brown stains. If found, iron is present.

3.4.2   Colorimetric method

(A) Supplier: Camlab Ltd.
Kit Description: HACH Iron Test Cube Kit. Cat. No. 14008 – 00.
Range: 0-5 mg/l in 1 mg/l increments. Sample size 5 ml.
No. of tests per kit: 50
'Ferrover' Powder Pillows.
Cost: £13.00.

Kit Descriptions: HACH Iron Test Kits, 3 kits to cover 3 ranges.
IR – 18A 0-1 mg/l, increments 0.02 mg/l.
IR – 18 0-5 mg/l, increments 0.1 mg/l.
IR – 18B 0-10 mg/l, increments 0.2 mg/l.
Sample size: 25 ml, 5 ml, 5 ml, respectively.
No. of tests per kit: 100.
Cost: IR – 18A   £45.50
      IR – 18   £26.50
      IR – 18B   £26.50
Note: Ferrover reagent is only stable for 6-12 months. Its effectiveness can be checked by testing water containing brown rust.

 If there are large amounts of iron (above 30 mg/l) in the sample the colour development may be inhibited. If this is suspected dilute the sample with deionised or distilled water to bring the value within the range.

 For most water testing purposes in developing countries the IR – 18 kit will suffice.

(B) Supplier: Wilkinson and Simpson Ltd.
Kit description: Lamotte Iron Test Kit P – 61 Code 4447.
Range: 0.5-10 mg/l in eight steps.
Cost: £21.45 for 100 tests.

Kit Description: New Palintest Lovibond Iron Test.
Range: 0-10 mg/l and 0-1 mg/l.
Lovibond Disc No: high range 3/117, low range 3/116.

Cost: Lovibond Disc: £19.00.
Cost: Lovibond 2000 Comparator: £22.80.
Cost: Palintest Thioglycolic acid regeant tablets, for high range per 250 in Al foil £4.75.
PPST reagent tablets per 250 bottle £5.45.

(C) Supplier: BDH Chemicals Ltd.
Kit description: 'Aquaquant' Iron Test Kits Cat. No. 165311W.
Range: 0.25-15 mg/l in 10 stage colour comparator Kit sufficient for 500 determinations.
Sample size: 10 ml.
Cost: £13.50.

3.5 Manganese

In natural waters manganese is usually found coexisting with iron, but is more of a problem because its oxide deposits are black, denser and often more difficult to remove. It is present in several oxidation states in surface water either as complexes or in suspension. In groundwater it usually exists as the Mn^{++} ion due to lack of oxygen but soon precipitates black, mixed oxides on exposure to air.

Levels greater than the WHO guide-level of 0.1 mg/l are sufficient to cause staining, washing and taste problems and for some industrial processes such as papermaking, dyeing and beverage manufacturing, the level must not exceed 0.02 mg/l.

The methods used to determine manganese are colorimetric based on oxidation to the purple permanganate. Tests should be made as soon as possible after sample collection.

Organically complexed manganese may not be totally oxidised in the periodate procedures and precipitated manganese may not be totally recovered.

Interferences
Cobalt and nickel may interfere by forming coloured complexes. Calcium and magnesium greater than 300 mg/l may give high readings.

3.5.1 Visual zero cost method (IDRC, 1981)
Equipment required: Clean white porcelain vessel (cup) or clean white paper.
Method: Pour fresh sample into cup and observe over a period of time. If a black or grey deposit or precipitate forms, manganese may be present.

Drip a few drops of sample onto white paper and allow them to dry. Check the edge of the watermark for black stains. If found, manganese may be present.

3.5.2 Colorimetric method

(A) Supplier: Camlab Ltd.
Kit description: HACH MN – S Manganese Test Kit Cat. No. 1467 – 00.
Range: 0-3 mg/l as manganese in 0.1 mg/l increments.
Sample size: 25 ml.
No. of tests per kit: 100.
Using 2 powder reagents in black plastic case.
Cost: £62.50.
Can also be obtained as combination kit with Iron Test Kit Model
IR – 20. Cat. No. 1463 – 00.
Cost: £79.50.
This test can be carried out using the DR – 100 hand held colorimeter.
The Manganese Kit plus colorimeter costs £180.00.

Kit description: Machery-Nagel 914018 Visocolor Kit Manganese
0.1-4 mg/l in steps 0.1, 0.3, 0.5, 1, 2, 4 mg/l. Kit contains 4 reagents,
3 of these are in plastic bottles one is a powder, a colour comparator
and compensation cell.
Sufficient for 200 tests.
Interference form iron if greater than 80 mg/l. Calcium and Magnesium
if greater than 300 mg/l.
Cost: £22.00.

(B) Supplier: Wilkinson and Simpson Ltd.
Kit description: Lamotte Model PMN Code 7518 Octet.
Comparator 0.2-3 mg/l with values 0.2, 0.4, 0.6, 0.8, 1.0, 1.4, 2.0,
3.0 mg/l steps. Persulphate reagent. Two reagents 10-15 minutes per
test. Kit sufficient for 50 tests.
Cost: (approx) £59.00.

3.6 Lead

Natural waters contain only slight traces of lead in solution because it is
preciptated by several substances. In groundwater its concentration is
normally' less than 10 μg/l or 0.01 mg/l. Higher values than 10 μg/l
indicates either:

> Lead dissolved from pipes and tanks painted with lead based
> paints, especially in soft water areas;

or

> Industrial mine or smelter wastes.

Lead compounds are poisonous; accumulation of significant
amounts of lead may cause severe and permanent brain damage or
death.

A concentration 50 μg/l or 0.05 mg/l of lead in drinking water is
considered the guide-level limit permissible (WHO, 1983). Regrettably

the determination of lead (at these low concentrations under field conditions) using dithizone is difficult. In hot countries these problems are aggravated by:

1. High volatility of organic solvents.
2. Higher temperatures and intense sunlight leading to rapid degradation of organic reagents and solvents.
3. Volume of solvents used is high.
4. Availability of lead-free chemicals.
5. Cyanide is used to complex other heavy metals.

It is better to find a suitable laboratory to undertake the determination of lead by the atomic absorption method or the dithizone method as described in APHA (1980).

There is a need for an appropriate field method to be developed but the methods described below can be used, with care, in the field to detect gross pollution but are not sufficiently accurate for regular monitoring near the WHO guideline level.

Interferences
Copper, tin, bismuth, chromates and sulphides interfere with the dithizone method.

Insoluble lead compounds are determined by the field methods but organo-lead compounds may only partially decompose in the digestion procedures involved.

3.6.1 Lead test kits
Supplier: Wilkinson and Simpson Ltd.
Kit description: Lamotte Lead Test Kit Model PPB Code 7662 Octet Comparator.
Uses modified APHA reagents with Dithizone.
Range: 0-1.5 mg/l.
Comparator increments: 0, 0.1, 0.2, 0.4, 0.6, 0.8, 1.0, 1.5 mg/l.
Five reagents sufficient for 50 tests.
Cost: £45.00.

Supplier: Camlab Ltd.
Kit description: Visocolor Kit Lead Cat. No. 914026.
Range: 0-1 mg/l. Increments: 0, 0.05, 0.1, 0.2, 0.5, 1.0 mg/l.
Dithizone reagents and equipped in two boxes.
Sufficient reagents for 50 tests.
Cost: £23.00.

3.7 Chromium
Chromium is a toxic element which occasionally occurs as a pollutant of natural waters due to its use in chromium plating processes, leather

51

tanning and as a corrosion inhibitor in industry. The hexavalent salts, chromates and dichromates, are the most common form of toxic chromium. Drinking water should not contain more than 0.05 mg/l of total chromium (WHO, 1983).

The colorimetric procedures rely on the sensitive reaction between hexavalent chromium and diphenylcarbazide producing a red-violet colour.

To determine total chromium, the chromium present in other forms should be converted to the hexavalent form by acid digestion and oxidation by potassium permanganate.

The detection limit of the field kits is above the WHO Guideline value and, as such, the kits are of limited value apart from monitoring gross pollution.

Interferences
The reaction with diphenylcarbazide is nearly specific for chromates. Heavy metals interfere as do all substances that colour the water or send it turbid.

3.7.1 Colorimetric methods
Supplier: Camlab Ltd.
Kit description: HACH Hexavalent Chromium Test Kit CH – 8 Cat. No.1834 – 00.
Range: 0 – 1.5 mg/l Cr. Increments 0.1 mg/l.
Sample size: 5 ml.
No. of tests per kit: 100.
Black plastic case.
Cost: £32.25.
Note: The reagents in this kit may be harmful. Avoid contact with eyes and skin. Do not swallow, and obey warnings on all chemical containers.
This method can be used in conjunction with the HACH DR – 100 hand-held colorimeter.

Kit description: Cr kit + Colorimeter Cat. No. 41100 – 03.
Range: 0 – 0.5 mg/l.
Cost: £180.00.

Kit description: HACH Chromium Test Cube Kit Cat. No. 12527 – 00.
Range: 0 – 1.0 mg/l.
Smallest increment 0.2 mg/l.
Sample size: 5 ml.
No. of tests per kit: 50.
Cost: £13.00.

Supplier: Wilkinson and Simpson Ltd.
Kit description: Palintest Chromate Test Kits CR – 114 Chromate.

Low range 0-100 mg/l Cr O₄ 20 tests.

Low range 0-100 mg/l Cr O_4 20 tests.
Single tablet reagent gives a colour change from red to colourless in low levels of chromate. For gross pollution only.
Cost: £6.75.

3.8 Zinc

Natural waters only rarely contain zinc but, because of the extensive use of galvanised tanks and brass fittings, water samples from distribution systems often contain small amounts of zinc. Above 5 mg/l of zinc the water may have a bitter astringent taste, opalescent appearance and even sand-like deposits of zinc carbonate. The solution of zinc from brass fittings and galvanised tanks may indicate that other metals such as lead and cadmium may also be present because they are impurities in the galvanising process (APHA, 1980).

Zinc in water can also be the result of industrial pollution.

Zinc is a trace element essential to plant and animal physiology, but because of the aesthetic problems such as taste and appearance the WHO Guide level is 5 mg/l on organoleptic grounds (WHO, 1983).

Zinc can be determined:

1. By its reaction with zincon at pH9. Other heavy metals are complexed by cyanide and sodium ascorbate whereas zinc-zincon complex is released by cyclohexanone. The reagents for this test are poisonous and care should be taken in disposing of testing chemicals.
2. By its reaction with Brilliant Green dye, with suitable buffers to complex iron, copper and other heavy metals.

3.8.1 Colorimetric tests

(A) Supplier: BDH Chemicals Ltd.
Kit description: Aquaquant Zinc Test Merck 14412.
Range: 0 – 5 mg/l in steps of 0.1, 0.2, 0.3, 0.4, 0.5, 0.7, 1.0, 2.0, 5 mg/l.
Four reagents sufficient for 120 tests.
5 ml sample size. In cardboard kit box.
Cost: £13.50.

(B) Supplier: Wilkinson and Simpson Ltd.
Kit description: Palintest WS3102 Zinc Comparator Kit using Lovibond Colorimeter.
Range: 0 – 4 mg/l Zn. Lovibond Colour Disc 3/102 Scale readings 0, 0.5, 1.0, 1.5, 2.0, 2.5, 3.0, 3.5, 4 mg/l Zn.
Using Palintest Zinc tablets. Sufficient for 100 tests.
Includes Lovibond 2000 Colorimeter.
Cost: £55.10.

(C) Supplier: Camlab Ltd.
Kit description: HACH Zinc Test Kit using DR – 100 Colorimeter.

Range: $0 - 3.0$ mg/l Zn and $0 - 1.5$ mg/l Zn using Zincon W/ZincoVer TM5.
Reagents: Sufficient for 100 tests. Cat. No. 41100 – 20.
Cost: £195.00.

3.9 Copper

Copper is not usually found in natural waters above trace values (greater than 0.01 mg/l). Copper salts are, however, used to control algae and plant growth in lakes and reservoirs, and in some developing countries to kill snails as part of bilharzia control programmes. Copper may be dissolved from corrosion of copper pipes and alloys and its presence may affect other metals in the distribution systems, especially galvanised and aluminium ware. As little as 0.1 mg/l copper may attack galvanised iron storage cisterns. Consumers often complain of green copper staining of baths and taps in the presence of alkaline soaps especially with hot water systems.

Copper is essential to human and plant physiology although large doses may cause stomach irritation and possible liver damage (Holden, 1970). Concentrations of over 1 mg/l will impart a bitter taste to water and WHO (1983) gives a Guide level of 1.0 mg/l on organoleptic grounds.

The colorimetric methods available for use in the field are based on reactions with complex reagents such as cuprizone (oxalic acid bis cyclohexylidene hydrazide), $2 - 2' -$ biquinoline $4 - 4'$ dicarboxylic acid and sodium diethyldithiocarbamate.

Interferences
Large amounts of chromium and tin may interfere. Cyanide, sulphide and organic matter interfere but can be removed by H_2SO_4/HNO_3 digestion procedure.

3.9.1 Colorimetric method
(A) Supplier: Camlab Ltd.
Kit description: HACH Copper Test Cube Kit Cat. No. 14497 – 00.
Range: $0 - 2.5$ mg/l as copper. Smallest increment 0.5 mg/l.
Sample size: 5 ml.
No. of tests per kit: 50.
Cost: £13.00.

Kit description: HACH Copper Test Kit CV – 5. Cat. No. 14213 – 00.
Range: $0 - 5$ mg/l Copper. Smallest increment 0.1 mg/l.
Colorimetric continuous disc.
Sample size: 5 ml.
No. of tests per kit: 100.
Black plastic case.
Cost: £32.25.

54

The DR – 100 hand held colorimeter may also be used with this method, the complete kit including colorimeter and reagent for 100 tests covering range 0 – 3 mg/l Copper, Cat. No. 41100 – 06. Cost £180.00.

(B) Supplier: BDH Chemicals.
Kit description: Aquaquant Copper Test Kit Cat. No. 165271 Y Merck 14414.
Range: 0.05 – 0.4 mg/l with 10 stage colour comparator and reagent for 150 determinations.
Cost: £13.50.

Kit description: Aquaquant Copper Test Kit Cat. No. 16528 K Merck 14418.
Range: 0.25 – 5 mg/l with 10 step comparator and Reagent for 150 determinations.
Cost: £13.50.

(C) Supplier: Wilkinson and Simpson Ltd.
Kit description: Palintest Comparator Kit WS3106 Copper LR Tablet Method. Range: 0 – 1 mg/L copper using Lovibond Comparator and Disc 3/106.
Reagents for 100 tests.
Cost: £55.10.

Kit description: Palintest Comparator Kit WS3110 Copper tablet method.
Range: 0.0 – 4.0 mg/l copper using Lovibond Comparator and Disc 3/110.
Reagents for 100 tests.
Cost: £55.10.

Kit description: Lamotte Model PCL Octet Comparator Kit 6616.
Range: 0.05 – 0.5 mg/l Copper.
Single reagent in powder form.
Sufficient for 50 tests.
Cost: £30.00.

3.10 Chloride

Chloride, as the chloride ion Cl^-, is a major constituent in water and waste water with a wide range of concentrations from a few milligrams per litre in clean rain to tens of grams per litre in supersaturated, hot, saline groundwaters. Chloride in rain chiefly originates from ocean spray and the concentration of chloride in rain at the coast is generally higher than at inland sites.

Chloride may be increased in surface water since it is concentrated in human and animal urine reaching water courses. Human urine may contain 1 – 1.5 per cent of sodium chloride. An increase in the chloride content (especially in the range 30 – 300 mg/l) in natural waters may be caused by pollution by sewage unless there are geological grounds for the increase.

Paper works, galvanising plants, softening plants and other industries may also discharge effluent containing chlorides, and even run-off from heavily fertilized fields can contribute chloride.

The origin of chloride in groundwaters is more complex since increased chloride values could be due to trapped fossil sea water, leaching of evaporite deposits or even washing out of chloride from overlying rocks and soils. In coastal regions overpumping can lead to increased chlorides in groundwater because of sea water intrusion. In areas of saline groundwater overpumping of some fresh water lenses above the saline layers may also induce the salt water to penetrate the system. If penetrated by deep drilling, artesian saline waters can leak into upper fresh water aquifers.

The effects of chloride on taste may be critical to its use. Some waters containing 250 mg/l chloride may have a detectable salty taste if the predominant cation is sodium but if the predominant cation is calcium or magnesium the salty taste may not be apparent even up to 1000 mg/l of chloride.

The Guide level of chloride according to WHO (1983) is 250 mg/l for drinking water on organoleptic considerations. Chlorides accelerate corrosion of concrete, cement and iron. Plants show variable sensitivity to chloride concentration. Grass, cotton and sugar beet are relatively resistant, but fruits and trees are more sensitive, with tolerance limits as follows (UNESCO, 1978):

Fruit	*Maximum tolerable*
Citrus fruits, grapes	700 mg/l Cl^-
Stone fruits, plums	500 mg/l Cl^-
Soft fruits, strawberries, raspberries	350 mg/l Cl^-

There are several methods used to determine chloride concentrations in the field namely the classic Mohr titration with Silver Nitrate for which drop titrations, the HACH digital titrator or Palintest tablets can be used, or the mercuric nitrate titration and the mercuric thiocyanate colorimetric method for which only drop titrations or the HACH digital titrator are suitable.

Interferences
Iodide and bromide are titrated with chloride. Chromate, ferric sulphate and phosphate ions in a concentration greater than 10 mg/l will interfere.

3.10.1 Tablet titration
Supplier: Wilkinson and Simpson Ltd.
Kit description: Palintest Chloride (Salinity) Test Kits.
Range: $0 - 1000$ mg/l Cl^- CS113 Compact Kit, 20 tests.
Cost: £7.50.

Kit description: SS213 Standard Kit, 100 tests.
Cost: £21.85.

3.10.2 Drop titration
Supplier: Camlab Ltd.
Kit description: HACH Chloride Test Kits.
Model $7 - P$ Cat. No. 1440. Drop Titration with Silver Nitrate.
Range: $0 - 2500$ mg/l NaCl. Sample sizes 23 ml and 5.83 ml.
One drop reagent $= 12.5$ mg/l and 50 mg/l respectively.
Number of tests (average): 100. Plastic case.
Cost: £17.10.

3.10.3 HACH digital titrator
Supplier: Camlab Ltd.
Kit description: HACH Model CD50 High Range Chloride Test Kit Cat.
No. $2086 - 00$.
Range: $1000 - 30\,000$ mg/l as NaCl.
Uses same principle as but stronger $Ag\ NO_3$.
1 drop $= 1000$ mg/l as NaCl.
No. of tests: 100.
Cost: £44.35.

Kit description: HACH High and Low Range Chloride Test Kit using
Digital Titrator $CD - DT$. Cat. No. $20635 - 00$.
Ranges: $0 - 100$ mg/l, $0 - 1000$ mg/l and $1000 - 20\,000$ mg/l as
Chloride Cl^-.
Smallest increment: 0.1 mg/l, 1 mg/l and 20 mg/l respectively.
Sample sizes: 100 ml,100 ml and 5 ml respectively.
No. of tests: 100 (50 low range, 50 high range).
Uses mercuric nitrate titration cartridges.
Titration flask, graduated cylinder and diphenyl carbazone indicator
powder pillows included.
Cost: £138.00

3.10.4 Colorimetric method
Supplier: BDH Chemicals Ltd.
Kit description: Aquaquant 14401 Chloride Test Kit.
Range: $5 - 500$ mg/l in 10 steps.
Sample size: 5 ml.

Colour comparator using Hg(SCN)$_2$ reagent (POISON).
50 tests in plastic kit.
Cost: £13.50.

3.10.5 Specific ion method
Supplier: MSE – Fisons Ltd.
Kit description: Orion model 94 – 17B chloride ion electrode for use with specific ion meter.
Range: 2 – 3800 mg/l chloride.
Cost: £179.00.

3.11 Sulphate
Sulphates occur in most natural waters in a wide range of concentrations. High values of sulphate above 200 mg/l can lead to attacks of diarrhoea especially in newcomers to a high sulphate water supply. After a few weeks' acclimatization the body becomes adjusted to higher values of sulphate. The WHO Guide level is 400 mg/l on organoleptic grounds. Waters in contact with sulphate rocks such as gypsum often have high sulphate values. Acid mine water, particularly from sulphide-bearing ores, and industrial wastes may also contribute large amounts of sulphate to natural waters.

In the beer brewing industry high sulphate waters are advantageous since they produce better flavours in the product.

Concrete is corroded by waters containing sulphate in quantities above 1000 mg/l and high sulphates can cause scaling on pipes, condensers and boilers. In developing countries drinking water containing high sulphates can contribute to problems of sewer corrosion.

The analysis of sulphate is difficult even in the laboratory and is subject to both positive and negative interference. The field methods are not very precise and either rely on the turbimetric precipitation of insoluble barium sulphate or EDTA complex titration in tablet form.

The test paper described relies on the discoloration of a thorin barium complex.

3.11.1 Test paper method
Supplier: BDH Chemicals Ltd.
Kit description: Merckoquant 10019 Sulphate Test. Transition intervals 0 – 200 mg/l, 300 – 400 mg/l, 500 – 800 mg/l, above 900 mg/l.
100 strips in aluminium tin.
Cost: £4.67.

3.11.2 Turbidimetric comparator
Supplier: Camlab Ltd.
Kit description: HACH SF – 1 Sulphate Test Kit Cat. No. 2251 – 00.

58

Range: 0.200 mg/l SO_4. Smallest increment 35 mg/l.
Sample size: 50 ml.
No. of tests per kit: 100. Black plastic case.
Cost: £31.25.

3.11.3 Palintest tablet kit
Supplier: Wilkinson and Simpson Ltd.
Kit description: Palintest Sulphate Test Kits. CS121 Sulphate low range.
0 – 200 mg/l. Compact kit about 20 tests.
Cost: £7.25.
SS121 Sulphate low range 0 – 200 mg/l. Standard Kit about 50 tests.
Cost: £20.25.
CS122 Sulphate High Range 0 – 2000 mg/l. Compact Kit about 20 tests.
Cost: £7.25.
SS122 Sulphate High Range 0 – 2000 mg/l. Standard Kit about 50 tests.
Cost: £20.25.
Note: Each Tablet No. 1 contains 0.038 grams of Barium Chloride which is POISONOUS. Do not swallow these tablets, and keep them secure.

3.12 Fluoride
The concentration of fluoride in drinking water is critical when considering the strength of growing teeth and bones.
Low levels of fluoride in the range 0.6 to 1.2 mg/l (depending on ambient temperatures) have been found to be effective in providing protection of tooth enamel in children and reducing dental caries (decay). Fortunately many natural waters have fluoride concentrations in this range. Waters low in fluoride sometimes have fluoride added to bring the concentration to the recommended level.
Staining of teeth, brittle bones and crippling in old people, may occur if the levels of fluoride are higher than about 2 mg/l. The WHO (1983) Guide level is 1.5 mg/l.
In areas of volcanic or igneous rocks, such as The Rift Valley in Africa, groundwaters may contain up to 20 mg/l of fluoride naturally. Fluorosis is inevitable in these areas if the water is used for drinking. Removal of fluorides from drinking water in developing countries by using bone charcoal, activated alumina or the Nalgonda process is not practised extensively.
Many governments restrict the use of high fluoride groundwaters and even close down expensively constructed wells due to high fluorides. There is a chance that these 'high' fluorides may only be apparently high because the methods of testing for fluorides colorimetrically are subject to several positive and negative interferences.

Recently the use of fluoride ion specific electrodes has improved the accuracy and reproducibility of fluoride determination to a satisfactory level. An economic decision on whether to close boreholes down due to an inaccurate method, or to invest in a specific ion electrode and meter, is often required.

Since the fluoride ion is one of the few chemical parameters that is not affected by transportation and storage (provided the sample is kept in a polythene bottle) the needs of fluoride analysis may be satisfied by having just one laboratory equipped with a fluoride-specific ion meter where samples may be sent and analysed.

The fluoride ion electrodes and meters described below are battery powered and could be used in the field.

The colorimetric method described uses the alizarin-zirconium reaction with tablet reagents and Nessler tubes (113 mm light path length). It is based on the fluoride ions decolorising a red violet colour to a pale yellow.

Interferences
Aluminium salts (greater than 0.1 mg/l), phosphates (greater than 5 mg/l) sulphates (greater than 200 mg/l) and high alkalinities and chlorides can cause positive and negative errors in the alizarin-zirconium method and SPADNS method.

With the fluoride ion method the TISAB buffer compensates for most interfering ions.

3.12.1   Colorimetric method
Supplier: Wilkinson and Simpson Ltd., and Lovibond-Tintometer Ltd.
Kit description: Fluoride test using Palintest A – Z tablets.
Range: 0 – 16 mg/l F in 0.2 mg/l steps.
Cost: Lovibond 2000 colorimeter £22.80.
Cost: Disc NOM (283738A) £19.00.
Cost: Nessler attachment DB412 £44.20.
Cost: 4(AF306) Nessler tubes (per pair) £4.55.
Cost: Palin Fluoride A – Z Tablets (100 in foil pack) £2.30.
Cost: Palin Fluoride "Excess AL" (250 in foil pack) £3.75.
Cost: Palin Fluoride Standard tablets (25) £4.00.

Supplier: Camlab Ltd.
Kit description: HACH DR/100 Portable Colorimeter in conjunction with SPADNS reagent.
Range: 0 – 2.0 mg/l increments 0.1 mg/l.
Reagent Powder Pillow F1212A. SPADNS.
Cost: DR/100 Colorimeter Kit = £215.00.
Four AA cell battery powdered or rechargeable.
This new development has yet to be tested by the author. However, it is still a colorimetric method and SPADNS reagent is not without interferences, although to a lesser extent than the alizarin method. The

60

SPADNS reaction is faster than the alizarin reaction and the colours can be read immediately.

If chlorine is present it must be removed by adding 1 drop of 0.5% sodium arsenite (Na AsO_2) solution per 0.1 mg/l chloride present before testing with SPADNS.

At present there are several portable colorimeters on the market which could be used instead of the DR – 100, but they would require standards to prepare a calibration graph.

At present the colorimeter DR – 100 from HACH can only be used with its own precalibrated scale for any one parameter. Previously, the HACH Engineers' Test Kits had interchangeable precalibrated scales so that one colorimeter could be used for many parameters. There may not be an easy way to change the wavelength filters on the DR – 100.

3.12.2   Specific ion electrode
Equipment required:
Supplier: MSE-Fisons Ltd.
Kit description: Orion 407 A/F Specific Ion Meter.
Cost: £775.00.
Orion Fluoride ion electrode. 94 – 09.
Cost: £227.00.
Orion Reference electrode 9001.
Cost: £59.00.
TISAB II Buffer solution.
Cost: £22.00 for 1 gallon.
Note that if TISAB II Buffer is to be prepared by a support laboratory it is important to obtain a supply of CDTA (1,2 Cyclo hexylene dinitrilo tetra acetic acid). This chemical is often difficult and expensive to obtain overseas. The HACH TISAB Buffer, supplied by Camlab (see below) is packaged in powder pillows and these could be used for other systems.

UK Supplier: Camlab Ltd.
Kit description: Fluoride Ion Analysis Package 13034 – 00 using the Model 16400 portable expanded scale pH meter.
Cost: £650.00.
Stock Fluoride Solution: (1 ml = 100 µg F).
Dissolve 0.221 ± 0.001 g of anhydrous sodium fluoride (NaF) in approximately 500 ml of deionised water. Dilute to 1000 ml in a graduated flask with distilled water. Mix well.
Total Ionic Strength Adjustment Buffer (TISAB): Place approximately 500 ml of distilled water in a 1-litre beaker adding 57 ± 0.2 Sodium Chloride and 4.0 g ± 0.1 g of CDTA (Cyclo Hexylene Dinitrilo Tetra Acetic Acid) and stir to dissolve. Place beaker in cool water bath and slowly and carefully add 6N sodium hydroxide solution until the pH is between 5.0 and 5.5 (usually about 120 ml of 6N NaOH is required).

Cool the solution and transfer to a 1-litre graduated flask. Dilute to the mark with deionised water and mix well.

Note: The temperature of all solutions should be the same since the response of electrodes is temperature dependent.

If problems occur refer to Instruction Manuals for Orion Fluoride Electrodes and Orion 407A meter.

3.13 Nitrate

Nitrate is an end-product of the decay of nitrogenous material such as nitrate fertilizers or animal and human excreta (Hutton and Lewis, 1980). In developing countries especially there is a risk of groundwater pollution by on-site sanitation (Lewis *et al.* 1982). High nitrate values in groundwater may serve as indicators of this type of pollution. Nitrogen fertilizers are causing high levels of nitrates in water supplies in some parts of the UK, and there are some cases in India of groundwaters and surface waters receiving high nitrate loads from fertilizers.

The level of nitrate in surface waters fluctuates with the seasons, influencing algae and plant growth rates which can degrade stream and lake water quality.

Samples of water require special storage techniques if analysed in a laboratory remote from their source, so field tests for nitrate give a more accurate reflection of the true nitrate concentration.

High nitrate concentrations in drinking water have been implicated in deaths of infants by cyanosis (metheamoglobinanemia) and possible cancer forming agents, though evidence is scanty and often confusing (Lewis *et al.* 1982). The recommendations from WHO (1977) were as follows:

	mg/l as NO_3^-	Category
General population	below 50	acceptable
	between 50 – 100	borderline
	over 100	unacceptable
Infants below age of 6 months	over 50	unacceptable

The WHO Guide Level is 10 mg/l as N or 44.3 as NO_3^- (WHO, 1983). It should be noted that some instruments report the nitrate as NO_3^- mg/l and others as N mg/l. The conversion factor below should help.

1 mg/l Nitrate as N = 4.43 mg/l nitrate as NO_3^-.

There are several methods available now for the rapid assessment of nitrate concentrations in the field using paper test strips, colorimeter kits and even nitrate-sensitive electrodes in conjunction with specific ion or expanded scale pH meters.

Interferences
Nitrites, Cyanides and oxidizing agents such as chlorine interfere with the colorimetric methods.

The nitrate ion electrode method suffers from interference from organic compounds, chlorides and bicarbonate ions, but buffer solutions can normally minimise the problems. However, the nitrate ion electrodes are subject to very small pH changes causing interference in their response which the buffer does not control adequately.

3.13.1   Test papers
Supplier: BDH Chemicals Ltd.
Kit description: 31524 Merckoquant Nitrate Test Strips 10020.
Box of 50 = £4.67.
For identification and semi-quantitative determination of nitrate ions.
Note: Nitrite NO_2^- may interfere and will be indicated by both sections turning purple. One drop of 10% solution of amidosulphonic acid per 1-ml of sample should be added to a sample for retesting to eliminate nitrite interference.
These strips are very useful for field work but when possible store them in a refrigerator and keep the top on the aluminium container. They are cheap and lightweight, but do have a limited shelf life and so fresh stocks should be obtained where possible. If the indicator strip is grey the tests will not be accurate.

3.13.2   Colorimeter
Supplier: Camlab Ltd.
Kit description: HACH N1 – 11 High Range Nitrate Test Kit.
Range: 0 – 50 mg/l as Nitrate as N or 0 – 222 mg/l Nitrate as NO_3^-.
Sample size: 5 ml.
No. of tests per kit: 100
Packed – black polypropylene case.
Cat. No. 1468 – 03.
Cost: £33.65.
This kit can also be obtained packed with the Nitrite Test Kit Model N1 – 12 at a cost of £50.00 for the two.

These have been used by the author in Botswana, Libya, Argentina and Sri Lanka. The powder pillow reagents come in boxes of 50 and are designed and tested to match the colour disc by HACH.

3.13.3   Specific ion electrode
Supplier: MSE-Fisons, MSE Scientific Instruments.
Kit description: Orion Specific Ion meter 407 A/F, Nitrate ion electrode (Orion Model 93 – 0&), Reference electrode (90 – 02 double junction) glassware, stirring rod.
Cost: £1107.00.

Although the cost of this equipment is high, it is a means of screening samples for nitrate in good quality groundwater. However, expertise and practise are necessary to use the isntrument properly and laboratory back-up will be needed for preparation of buffers and standards.

Reagents:
Stock nitrate solution 100 mg/l nitrate as NO_3^-.
Dissolve 3, 1965 g of anhydrous potassium nitrate KNO_3 and dilute to 1000 ml with distilled water in a volumetric flask. Stabilise for about 2 months, discard if any algae growth appears. Store in the dark.
ISA Buffer: 2M Ammonium sulphate
26.4g $(NH_4)_2 SO_4$ in 100 ml volumetric flask. Dissolve and dilute to mark.

3.14 Nitrite
Nitrite is an intermediate product of the decay of nitrogeneous material. It is formed by the bacterial oxidation of ammonia and then rapidly oxidised by bacterial action to nitrate. Its presence in a water supply usually denotes bacterial activity as a result of recent or on-going pollution, often from sewage. On the spot field testing will assist in tracking back to the source of pollution. Even in polluted waters, nitrite concentration is usually not greater than 2 mg/l.

Nitrites are often used as corrosion inhibitors in process waters for industry and are extensively used in meat preservation.

There are two means of testing in the field for nitrite: test papers and comparator kits.

3.14.1 Test papers
Supplier: BDH Chemicals Ltd.
Kit description: 31524 'Merckoquant' Nitrite Test (10007) for the detection and semi-quantitative determination of nitrite ions.
Range: 0 – 50 mg/l in steps 0, 1, 5, 10, 25, 50 mg/l.
Cost: of aluminium container of 100 strips £4.67.
Note: Store unopened containers in the refrigerator. Opened containers should be stored in a cool DRY place but not in a refrigerator (to prevent condensation).
Note: It would be more useful to have a test paper to cover range 0 – 2 mg/l.

3.14.2 Colorimeters
Supplier: Camlab Ltd.
Kit description: HACH Nitrite Test Cube Kit 20596 – 00.
Range: 0 – 1 mg/l in 0.2 mg/l increments 5 ml sample size. In case, with 50 Nitriver test powder pillows.
Cost: £12.00:

HACH Nitrite/Nitrite Test Kit N1 – 12 (see Nitrate tests).
Range: 0 – 1.65 mg/l as Nitrite NO_2^- in 0.03 mg/l increments.
Cost: £50.00.

Supplier: Lovibond Tintometer Ltd.
Kit description: Lovibond 2000 colorimeter and Palintest Nitricol tablets.
Lovibond Disc 3/103.
Range: 0.05, 0.1, 0.15, 0.2, 0.25, 0.3, 0.35, 0.4, 0.5, mg/l as Nitrite.
Cost: Lovibond 2000 £22.80.
250 Nitricol tablets in Aluminium foil £4.75.
Disc 3/103 £19.00.

Supplier: BDH Chemicals Ltd.
Kit description: 165321 Aquaquant Nitrite Test Kit
Range: 0 – 1 mg/l Nitrite about 200 test. In cardboard pack.
Cost: £15.00.

UK Supplier: Camlab Ltd.
Kit description: Machery-Nagel Visocolor Kit Nitrite.
Range: 0.05 – 2 mg/l cube comparator.
Sufficient for 50 tests. Two reagents in dropper bottles. Plastic case.
Cost: £22.00.

3.15 Ammonia

Trace amounts of ammonia are found in most natural waters. Sewage contains large amounts of ammonia formed by bacterial decay of nitrogenous organic wastes. Hence surface waters showing a sudden increase in the ammonia content may indicate sewage pollution or industrial pollution from dairies, abattoirs, tanneries or chemical plants. Groundwaters often contain some ammonia due to natural processes from reduction of nitrates by bacteria or by well-construction materials, but sudden changes in ammonia content need further investigation.

The nature of ammonia in solution is pH dependent; below pH 6 the ammonia is combined as the ammonium NH_4^+ ion, above pH 7 the percentage of free ammonia increases rapidly. See table below.

pH	% Free ammonia
6	0
7	1
8	4
9	25
10	78
11	96

Free ammonia will increase the chlorine demand of a raw water in the chlorination process of disinfection. Its presence is not proof of contamination but may provide supporting evidence of pollution. It may also be an index of the progress of self-purification taking place in a river.

All the older field tests rely on the Nessler's reagents, in strongly alkaline conditions, reaction with the ammonia present in the sample to produce a yellow-coloured product. Measurement of the yellow colour gives an accurate method of determining the ammonia concentration.

It should be noted that Nessler reagent contains mercury and is TOXIC and must not be swallowed. After testing the coloured solutions should be disposed of away from possible sources of drinking water or food.

In saline waters and those containing high levels of magnesium, precipitates may interfere. This can be alleviated by adding 2 drops of 50% Richelle salt (Na K C_4 H_4 O_6. $4H_2$ O, Sodium potassium tartrate).

The field methods available are essentially colorimetric although one test strip method can be used. Recently, however, the phenate method has been adapted by Palintest (1982) for field use. It uses stable tablet reagents and (unlike the Nessler method) is more sensitive to low levels of ammonia and does not use reagents containing TOXIC mercury compounds. This method is therefore preferable.

3.15.1 Test paper method
Supplier: BDH Chemicals Ltd.
Kit description: 31526 Merckquant Ammonia Test Merck 10024.
Range: 0 – 400 mg/l in steps 0, 10, 30, 60, 100, 200, 400 mg/l.
Cost: £5 for 50 tests.
Storage: one year maximum.
Probably not accurate enough for water quality work.

3.15.2 Comparator methods
(A) Supplier: BDH Chemicals Ltd.
Kit description: 165241C 'Aquaquant' Ammonia (NH_4^+) Test Kit Merck (14400).
Range: 0.05 – 0.8 mg/l.
Sufficient for 100 determinations.
Cost: £13.50.

Kit description: Aquamerck Ammonia Test Merck 1117.
Reagent kit with cardboard colour scale 0, 0.5, 1.0, 3.0, 5.0, 10 mg/l NH_4^+ with 3 reagent dropping bottles sufficient for 150 determinations.
Cost: £22.00.

(B) Supplier: Camlab Ltd.
Kit description: Macherey Nagel Cat. No. 914010 Visocolor Kit Ammonia.

66

Range: 0.2 – 10 mg/l NH_4^+ Colour cube with scale 0.2, 0.5, 1.0, 2.0, 5.0, 10 mg/l NH_4^+.
Sufficient reagents for 150 tests. Includes comparator cube, 2 x 50 ml of reagents, 2 dropping pipettes in plastic box.
Cost: £22.00.

Kit description: HACH Test Cube Kit 12524 – 00.
Range: 0 – 2.5 mg/l in 0.5 mg/l increments.
Sample size: 5 ml.
Sufficient reagent in bottle for 100 tests.
Cost: £12.00.

Kit description: HACH Ammonia Nitrogen Test Kit N1 – 8 Cat. No. 2241 – 00. Use continuous colour disc comparator.
Range: 0 – 3 mg/l, smallest increment 0.1 mg/l.
Sample size: 5 ml.
Sufficient reagent for 100 tests. Rochelle Salt reagent to overcome interference available separately. (Cat. No. 1725 – 23)
Cost: £27.50.

(C) Suppliers: Wilkinson and Simpson Ltd.
Kit description: Palintest Ammonia Test for use with Lovibond 2000 Comparator (or Lamotte MTRL Colorimeter).
Range: 0 – 0.4 mg/l as NH_4 or 0 – 1 mg/l Nitrogen.
Cost: Lovibond comparator 2000 = £22.80.
Palintest Ammonia tablet per 250 in foil packs.
Cost: Ammonia No. 1 £3.55.
Cost: Ammonia No. 2 £3.55.
Lovibond Colour Disc No 3/112 for use with 40 mm cells (Range 0 – 0.4 mg/l) in steps of 0.05 mg/l (20 ml sample NH_4).
Cost: £19.00.
Lovibond Colour Disc No 3/133 for use with normal 13.5 mm cell (10 ml sample).
Range: 0.0 to 1.0 mg/l Nitrogen. 0 to 1.3 mg/l NH_4.
Cost: £19.00.
Lovibond 2000 Colorimeter.
Cost: £22.00.
Note: Concentrations of ammonia greater than the highest standard on the disc may be determined by diluting the sample with ammonia-free distilled or deionised water. The reading obtained is then multiplied by the dilution factor.

Notes:
1.   Where the sample temperature is below 20 °C allow 15 minutes for maximum colour development.

2. The following factors may be used for conversion of readings:
   — from NH_4 to NH_3 0.94
   — from NH_4 to N   0.78
   — from N to NH_4 1.3

3.16 Phosphate

Phosphate commonly occurs in natural waters and is often added in water treatment chemicals. It exists principally in three states:
   orthophosphate, PO_4^{---},
   meta (or polyphosphate), PO_3^-,
   and organically bound phosphates.
Excessive amounts of phosphate actually constitute pollution, usually by infiltration of wastewaters from domestic and industrial sources or agricultural run off. Metaphosphates derived from detergents, hardness treatment systems and boiler waters slowly convert to orthophosphate.

Phosphates are often the limiting nutrients for growth of many organisms in water and too much phosphate can lead to rapid eutrophication especially in lakes, reservoirs and ponds where other nutrients such as nitrates may be present. Such rapid growth in hot climates where the dissolved oxygen in waters is already low can create problems of taste and odour.

The determination of orthophosphates depends on the reactions with molybdate in acidic conditions to produce phosphomolybdic acid which is then reduced to a heteropoly blue compound. The phosphate concentration of the sample is then determined accurately by measuring the intensity of the blue colour.

The metaphosphate must be hydrolysed to orthophosphate before it is determined and organic phosphates must be oxidised and hydrolysed before being determined as orthophosphate.

Silica and silicates greater than 10 mg/l can interfere with this determination and give high readings.

3.16.1 Colorimetric method

(A) Supplier: BDH Chemicals Ltd.
Kit description: Aquamerck Phosphate Test Kit 8046.
Range: $1.3 - 13.4$ mg/l PO_4^{---} in 1.3, 3.3, 6.7, 10, 13.3 mg/l.
Sample size: 10 ml.
No. of tests per kit: 100.
Cost: £6.71.
Note: orthophosphate PO_4^{---} only.

(B) Supplier: Camlab Ltd.
Kit description: HACH Total Phosphate Test Kit.

Model: PO – 24. Cat. No. 2250 – 01 for determination of ortho, meta and organic phosphates.
Ranges: 0 – 1 mg/l, 0 – 5 mg/l and 0 – 50 mg/l as PO_4^{---}.
Smallest increments: 0.02 mg/l, 0.1 mg/l and 1 mg/l respectively.
Sample size: 20 ml, 5 ml and 2 ml respectively.
No. of tests per kit: 100.
In plastic case.
Cost: £98.75.

Kit description: HACH Phosphate Test Cube Kit. Cat. No. 12522 – 00 for orthophosphate in clear waters.
Range: 0 – 5 mg/l as PO_4^{---}.
Smallest increment: 1 mg/l.
Sample size: 5 ml.
No. of tests per kit: 50.
Cost: £13.00.

Kit description: Visocolor Kit Phosphate 1 – 15 mg/l. Cat. No. 914023.
Range: 0 – 15 mg/l PO_4^{---}.
Increments: 1, 2, 5, 7, 10, 15 mg/l.
Sample size: 9 ml orthophosphate only.
No. of tests per kit: 100.
Cost: £23.00.

(C) Supplier: Wilkinson and Simpson Ltd, Lovibond Ltd.
Kit description: AF 620. Phosphate Kit.
Range: 0 – 100 mg/l PO_4^{---};
Increments: 0, 10, 20, 30, 40, 50, 60, 80, 100 mg/l.
Sample size: 10 ml.
No. of tests per kit: 100 orthophosphate only.
Uses Palintest tableted reagents (£5.05 per 100) includes Lovibond 2000 Comparator (£28.00). 3/70 disc (£19.00) in wooden box.
Cost per kit: £67.80.

3.17 Cyanide

Cyanide is not usually found in natural waters and its presence indicates pollution from an industrial source. Industries which pose a threat from dumping of cyanide wastes are metal cleaning and plating, gas works and coke ovens, steel works and other chemical plants using cyanide. Cyanide can be easily removed from these effluents by alkaline or neutral chlorination processes.

It is very TOXIC but often combines with metals to form less toxic complexes. The WHO Guide Level is 0.1 mg/l (WHO, 1983).

The field test relies on the pyridine-pyrozolone method which after a series of reactions produces a deep blue colour. Cyanide metal complexes do not react and interference from heavy metals may invalidate the test as a field procedure.

3.17.1 Colorimetric method

(A) Supplier: Camlab Ltd.

Kit description: HACH Cyanide Test Kit CYB – 3. Cat. No. 2010 – 02.

Range: 0 – 0.2 mg/l as CN⁻; smallest increment 0.01 mg/l.

Sample size: 5 ml.

No. of tests per kit: 100.

In black plastic case with instructions.

Cost: £64.90.

This test is also available for use with the DR100 hand-held colorimeter, Cat. No. 41100 – 07.

Cost: £180.00.

(B) Supplier: BDH Chemicals Ltd.

Kit description: 165291M 'Aquaquant' Cyanide Test Kit 14417.

Range: 0.002 – 0.03 mg/l in steps with comparator.

Reagent kit for 65 determinations.

Cost: £13.50.

Samples may need to be diluted to bring them into this low range.

3.18 Arsenic

Chronic effects of arsenic poisoning can appear from its accumulation in the body even at low concentrations. It is also suspected of having carcinogenic properties. It occurs in some groundwaters naturally, in Argentina for example there are some wells with 4.0 mg/l arsenic and some mineral waters contain higher levels. Normally groundwaters contain less than 0.1 mg/l arsenic. The WHO Guide level is 0.05 mg/l (WHO, 1983).

Natural waters may be contaminated by arsenic from refuse tip leachate, arsenic based weedkillers and insecticides or industrial discharges.

The field test for arsenic uses the mercuric bromide stain method. Arsine gas (As H_3) is generated from trivalent and pentavalent arsenic compounds reacting with zinc and hydrochloric acid. The generated arsine produces a yellow brown stain on the mercuric bromide impregnated paper. The length of stain is then compared to a scale printed on the side of the test paper container.

Organo-arsenic compounds may not decompose completely in the digestion process.

The field test low limit of detection is above the WHO Guide level and as such the kit should only be used to detect cases of gross pollution. Samples should be taken for more accurate determinations (see Table 3.1).

Interferences

Heavy metals, selenium and sulphides at low levels may interfere

70

causing colour reactions on the test strip, and antimony greater than 0.1 mg/l may give a false reading.

Supplier: BDH Chemicals Ltd.
Kit description: Merckoquant 10026 Arsenic (As^{3+}/$^{5+}$) Test.
Test strips, reagents, reactor and accessories for the detection and semi-quantitative determination of arsenic.
Colour scale graduated 0, 0.1, 0.5, 1.0, 1.7, 3.0 mg/l As.
Sample size: 5 ml.
Sufficient reagents for 100 determinations.
Cost: £12.26.

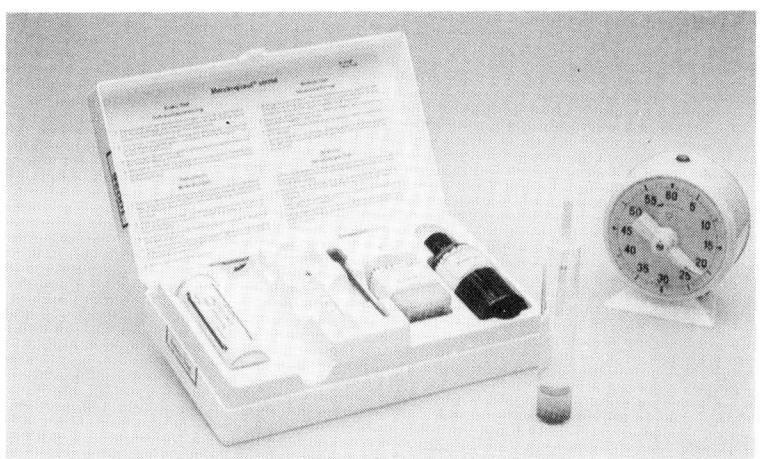

Figure 3.2   Merckoquant arsenic test

3.19   Alkalinity

The alkalinity of a water is caused by the presence of carbonates, bicarbonates and hydroxides. In most natural waters the principal anion is bicarbonate with carbonate and/or hydroxyl ions being present only if the pH is greater than 8.3 (Phenolphthalein alkalinity, P). Phenolphthalein alkalinity is usually due to carbonates and/or free hydroxides from industrial pollution in surface waters. Most surface waters show no phenolphthalein alkalinity, since all their alkalinity is due to the bicarbonate of alkali earth and iron. The relationship between pH and the presence of hydroxides, carbonates, bicarbonates and carbon dioxides is shown overleaf in Fig. 3.3.

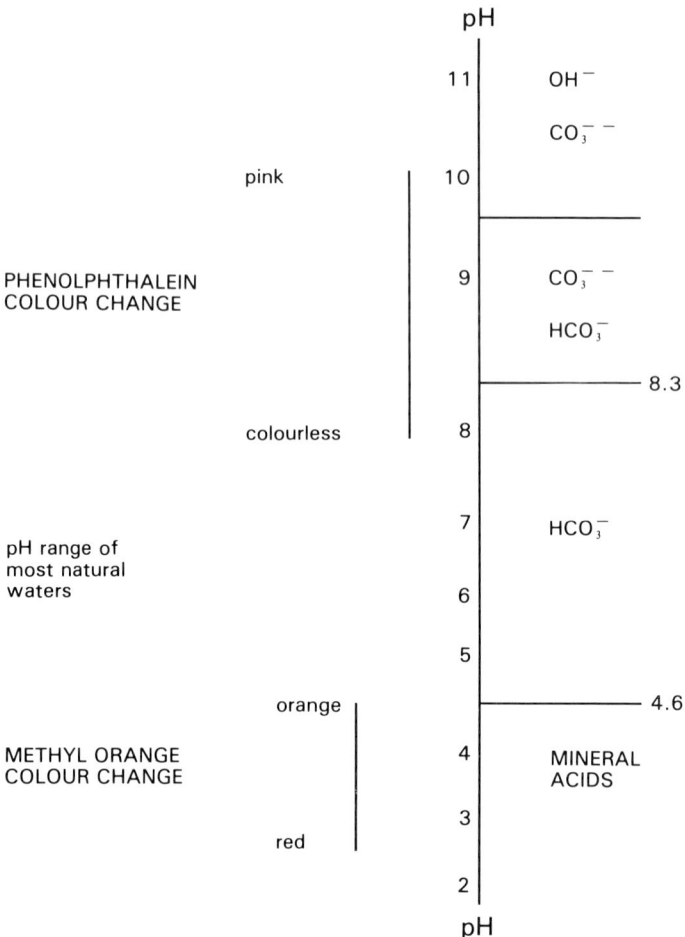

pH

11   OH$^-$

CO$_3^{--}$

pink   10

PHENOLPHTHALEIN   9   CO$_3^{--}$
COLOUR CHANGE

HCO$_3^-$

— 8.3

colourless   8

7   HCO$_3^-$

pH range of
most natural
waters   6

5

orange   — 4.6

METHYL ORANGE   4   MINERAL
COLOUR CHANGE   ACIDS

3

red

2

pH

Figure 3.3   Ion species and pH

Alkalinity is determined by titration of the sample against a known strength acid to the phenolphthalein end point pH 8.3 (pink to colourless) and then to the methyl orange end point pH 4.8 (orange to pink).

The procedure is to add a few drops of phenolphthalein indicator to 100 ml of the sample. If hydroxide or carbonate ions are present a pink colour will be seen. Known strength acid (usually sulphuric acid) is then added until this pink colour just disappears (pH 8.2). The amount of acid used is recorded as the *P value.*

Methyl orange indicator is then added to the same sample and titration with known strength sulphuric acid is continued until the colour changes from yellow to orange (pH 4.0). The quantity of acid for this operation is recorded as the *M value.*

The total volume of acid used (P + M) is known as the *T value* or Total Alkalinity Value.

There are 5 conditions of alkalinity in water which are recognised:

1. Hydroxide alkalinity when P = T. The alkalinity is due solely to OH^- ions and the acid is then equivalent to OH^- concentration ONLY.

2. Hydroxide and carbonate alkalinity, when P is greater than ½T but less than T. M is greater than zero.

The hydroxide (OH^-) concentration is equivalent to 2P – T ml of acid used.

The carbonate (CO_3^{--}) concentration is equivalent to 2T – P ml of acid used.

3. Carbonate alkalinity, when P equals exactly ½T, or P = M. In this case only carbonate is present.

Carbonate (CO_3^{--}) concentration is equivalent to 2P ml of acid used.

4. Carbonates and bicarbonates present, when P is less than ½T. Carbonate concentration is equivalent to 2P ml of acid used. Bicarbonate concentration is equivalent to (T – 2P) ml of acid used.

5. Bicarbonates only, no pink with phenolphthalein. Bicarbonate concentration is equivalent to T ml of acid used.

In order to calculate the various components of alkalinity the equivalents below may be used:

1 ml of Molar H_2SO_4 $\equiv$ 1 ml of 2N H_2SO_4 $\equiv$ 122 mg H CO_3^-
                             $\equiv$ 60 mg CO_3^{--} $\equiv$ 34 mg OH^-
                             $\equiv$ 100 mg Ca CO_3

Bromocresol green – methyl red indicator may be used instead of methyl orange especially for samples with various alkalinities and compositions below:

Sample variant	End point pH	Bromocresol Green Methyl red colour
Alkalinity about 30 mg/l Ca CO_3	5.1	Light green — blue grey
Alkalinity about 150 mg/l Ca CO_3	4.8	Light blue — pink grey
Alkalinity about 500 mg/l Ca CO_3	4.5	Light pink
Silicates or Phosphates present	4.5	Light pink

Most of the field methods report the alkalinity as calcium carbonate mg/l. For many applications it is preferable to convert the various alkalinities to their actual ionic concentrations using the equivalents described previously.

3.19.1 Tablet titration

Supplier: Wilkinson and Simpson Ltd.
Kit description: Palintest Alkalinity Test Kits
Alkalinity (Phenolphthalein) P CS 110 Compact Kit 20 Test
SS 210 Standard Kit 100 Tests.
Range: 0 – 500 mg/l Ca CO_3.
Cost: CS 110   £6.75.
        SS 210 £21.25.
Alkalinity M (Total) CS 112 Compact Kit 20 Tests.
SS 212 Standard Kit 100 Tests.
Range; 0 – 500 mg/l Ca Co_3.
Cost: CS 112   £6.75
        SS 212 £21.25
Also combination test kit.
SS 230 Alkalinity P/Alkalinity M 100 tests.
Range: 0 – 500 mg/l Ca CO_3.
Cost: £21.25.

3.19.2   Titration method

Supplier: Camlab Ltd.
Kit description: HACH Alkalinity Test Kit AL – AP. Cat. No. 1433 – 00.
For methyl orange (Total) and phenolphthalein alkalinity using drop count titration.
1 drop = 7 mg/l Ca CO_3 or 17 mg/l Ca CO_3 alkalinity as calcium carbonate.
Sample size: 15 ml and 5.83 ml.
No. of average tests: 50.
Cost: £16.75.

Kit description: HACH high and low range Alkalinity Test Kit with Digital Titrator Model AL – DT. Cat. No. 20637 – 00.
Ranges: 0 – 100 and 0 – 1000 mg/l as Ca CO_3.
Smallest increment: 0.1 and 1 mg/l respectively.
Sample size: 100 ml.
No. of average tests: 200 (100 P and 100M).
Cost: £130.00.

3.20   Acidity

Acidity of a water is its quantitative capacity to react with a strong base to a designated pH (APHA 1980). It is the opposite property of alkalinity. Fortunately, the presence of free mineral acids in water is not common and is mainly due to industrial pollution, but there can be natural causes such as humic acids in swamp or peat and sulphuric acid from oxidation of natural sulphides in mine waters. Following desalination of sea water by distillation in the Middle East, the distillate is often so acid that its neutralisation by chalk or lime is necessary to prevent rapid and severe injury to the teeth of consumers and severe corrosion of pipework and storage tanks. Acid waters may also contribute to solution of lead pipes giving rise to high lead levels.

Recently, acid rain from sulphur dioxide pollution has been shown to have severe effects on forests and lakes in Scandinavia and Northern Europe. In many cases biological cycles have been affected such that few fish or plants remain in large areas of surface water.

Field procedures for determining acidity involve titrating sodium hydroxide to the methyl orange (or bromophenol blue) pH 3.7 endpoint and then the phenolphthalein pH 8.3 endpoint to give acidity to methyl orange and total acidity to phenolphthalein.

The end points may be affected by the presence of iron, aluminium and manganese precipitating near the end points. Residual chlorine may bleach the indicator end point and should be removed by 1 drop of 0.1N Sodium Thiosulphate ($Na_2S_2O_3$) before titrating.

The pH of the end point should be reported and the acidity is expressed as acid equivalents of calcium carbonate.

Samples should be taken avoiding bubbling as described in carbon dioxide sampling procedures (Section 3.23).

3.20.1 Titration method
(A) Supplier: Camlab Ltd.
Kit description: HACH Acidity Test Kit Model AC − 5. Cat. No. 2223 − 00. Drop count titration with sodium hydroxide. Read out 1 drop = 7 mg/l or 17 mg/l acidity as calcium carbonate.
Sample size: 15 ml and 5.85 ml respectively.
No. of average tests: 100.
In plastic case.
Cost: £23.00.

Kit description: HACH Acidity Test Kit, high and low range with Digital Titrator Model AC − DT. Cat. No. 20640 − 00.
Ranges: 0 − 100 mg/l and 0 − 1000 mg/l acidity as Calcium Carbonate.
Smallest increment: 0.1 mg/l and 1 mg/l.
Sample size: 100 ml.
No. of average tests: 200.
In black plastic case.
Cost: £130.00.

(B) Supplier: Wilkinson and Simpson Ltd.
Kit description: Lamotte Model MB − FMA Dropcount Acidity Test Kit, Code 7604.

Range: 0 – 1000 mg/l, 1 drop = 5 mg/l. Four reagents.
No. of tests per kit: 25.
Cost: £16.00.

3.21 Chlorine

The presence of chlorine in drinking water is used as a measure of the effectiveness of chlorine as a disinfecting agent. Chlorine gas or chlorine compounds are very strong oxidising agents and kill bacteria in water. They also react with organic matter, ammonia, iron, and manganese. Sufficient chlorine needs to be added to water to carry out the above reactions and leave a slight excess to ensure complete disinfection.

Chlorine is applied to water either directly from gas cylinders or by use of bleaching powder or bleach containing amounts of free chlorine. Normally, bleaching powder or chloride of lime ($CaO. 2Ca OCl_2. 3H_2O$) contains between 37% – 25% available chlorine depending on its age and storage conditions. HTH, high test hypochlorite, $CA(OCl)_2 4H_2O$, is becoming more widely available and has 70% available chlorine when new. Household bleach solution ($Na OCl$) usually contains 5% – 15% available chlorine.

The reactions of chlorine and its compounds in water are described in Cox (1969) but the graph below summarises most of the reactions:

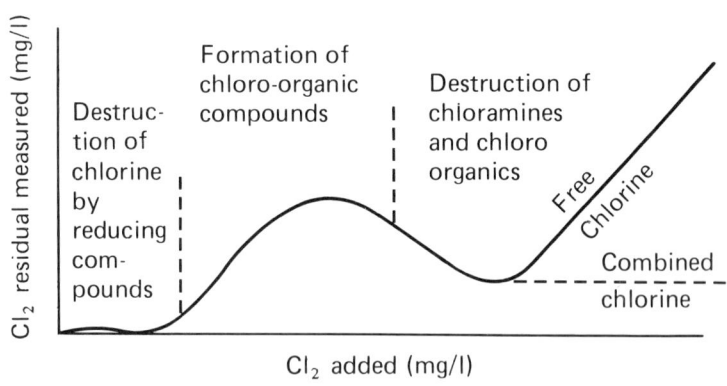

Figure 3.4   Reaction of chlorine in water

When chlorination is carried out in developing countries, the most important parameter to be determined is the free residual chlorine content. If we find free residual chlorine in a sample we may assume that, provided sufficient contact time between chlorine and the water has been allowed, the water will be bacteriologically safe at the point the sample is taken. This is no guarantee that contamination has not occurred elsewhere in the distribution system. A small amount of free residual chlorine 0.2 – 0.5 mg/l is unlikely to be strong enough to kill any bacteria invading via leaks or cross connections.

The detection of free residual chlorine therefore needs to be checked in all parts of the system to ensure that safe water is being distributed.

There are several reagents that have been used to determine chlorine residuals namely:

1. Ortho-tolidine
2. DPD (Diethyl-para-phenylene diamine)
3. Starch – potassium iodide.

Ortho-tolidine, however, has been found to be CARCINOGENIC (cancer inducing) and its sale and manufacture has been controlled by legislation. It is likely to be difficult to buy in future and for descriptions of its use the reader is referred to Cox (1969) or McJunkin (1976).

The methods described will cover the use of DPD and Starch/KI. The apparatus available for chlorine determinations is varied both in cost and sophistication but field tests depend on colour comparisons.

3.21.1   Colorimetric DPD chlorine testing
3.21.1.1   Chinese method (visual/nasal)(IDRC, 1981)
Equipment required: 5 ml sample bottle, dropper. DPD solution – 0.1 g DPD dissolved in 10 ml conc. hydrochloric acid and 90 ml distilled water.
Cost: Cheap.
Procedure: Add 3-5 drops of DPD solution to a 5 ml sample in a clear sample bottle. Yellow colour indicates residual chlorine according to table below:

Residual chlorine mg/l	Smell of chlorine	Colour
0.1	Slight	Light yellow
0.2	Easily detectable	Yellow
0.5	Obvious	Yellow
0.7 – 1.0	Strong	Deep yellow
greater than 20	Very strong	Brown

This method is fairly 'rough and ready' so more sophisticated techniques will be described.

3.21.1.2 Lovibond Comparator (based on Palin (1957)(from Thomas and Chamberlain (1980))

(A) Suppliers: Lovibond-Tintometer Ltd. and Wilkinson and Simpson Ltd.

Kit description: Lovibond Comparator 2000, 13.5 mm moulded cells, DB424, and discs to cover range expected.

Disc Code	Range Covered in mg/l Cl_2
3/40A	0.1, 0.2, 0.3, 0.4, 0.5, 0.6, 0.7, 0.8, 1.0
3/405	1.0, 1.2, 1.4, 1.6, 1.8, 2.0, 3.0, 4.0
3/40B	0.2, 0.4, 0.6, 1.0, 1.5, 2.0, 2.5, 3.0, 4.0
DPD Tablets No. 1	(for free residual chlorine)
DPD Tablets No. 3	(for free and residual chlorine)

These are available in aluminium foil strip packs in 'rapid dissolving' or standard tablets. Since there is no difference in cost the rapid dissolving form are recommended.

The shelf life of DPD tablets is very long. There are cases of DPD tablets being perfect after 10 years (Palintest, 1982).

There is a fear that DPD tablets may be eaten by children or older people thinking they were sweets or medicine. The low concentration of DPD in the tablets has not yet been known to be toxic even if large numbers of tablets are taken. However, it is suggested that aluminium foil packs of all types or tablets be kept away from children if possible.

Cost:

Lovibond Comparator 2000 (+ 2 tubes) £22.80

13.5 mm square test tubes (each) £1.50

Discs (each) £19.00

DPD Tablets: No.1 foil strips—100 (£1.90), 250 (£3.75), 1000 (£14.40) bottles

No. 3 foil strips—250 (£2.85)

Procedure:

(a) Free residual chlorine:

1.   Place in the left-hand compartment of the comparator, behind the colour standards of the disc, a 13.5 mm cell containing 10 ml of sample only (BLANK).

2.   Rinse a similar cell 3 times with the sample and then dissolve one No. 1 DPD tablet in a few drops of sample in the tube.

3.   After disintegration of the No. 1 DPD tablet make up the tube to the 10 ml graduation mark, mix sample thoroughly and place alongside the other tube in the comparator.

4.   Match the colours in the two tubes IMMEDIATELY, holding the comparator to north daylight or a standard source of light (not fluorescent tubes) and revolving the disc until a colour match is

78

Figure 3.5   Lovibond Comparator 2000

obtained. The figure shown in the indicator window, bottom right represents the free residual chlorine in mg/l.

(b) Combined residual chlorine:
1.   Measure the free residual chlorine as outlined in step 1-4 above.
2.   Add one DPD tablet No. 3 to the sample tube, mix and stand for two minutes.
3.   Compare the colours as in step A4 above, and read off the value indicated which represents the total residual chlorine.
4.   To obtain the combined residual chlorine deduct the free chlorine residual from step A4 from the total residual chlorine in step B3.
Combined residual Cl_2 mg/l = Total residual Cl_2 mg/l (B3) minus Free residual Cl_2 mg/l (A4).

(B) Supplier: Camlab Ltd.
HACH Chlorine Testing Kits:
   HACH produce several kits for testing chlorine in drinking water and their cost varies according to the accuracy and sensitivity of testing required. The reagents are used to indicate chlorine concentrations in the two types described below but are packaged in premeasured sealed plastic powder pillows opened by nail clippers.
   The discs used in HACH kits are made of plastic and when not in use should be protected from direct sunlight by putting them into the black protective packets provided to prevent colour deterioration by ultra-

violet and scratching by dust particles. One advantage of the plastic colour discs is that the colours are continuously increasing around the disc which makes for easier reading compared to the staged readings of glass standards.

Free and total chlorine (EPA approved):
Equipment required:
Kit description: HACH Test Kit CN – 66 – comes complete with comparator, colour disc, clippers and DPD reagent powder pillows for 50 tests free chlorine and 50 tests of total chlorine.
Range: 0-3 mg/l in 0.1 mg/l increments.
Cost: HACH Kit CN – 66 with reagents for 50 free Cl_2 and 50 total CL_2
    tests £28.25
    Reagent powder pillows (50 pillows) £5.00.

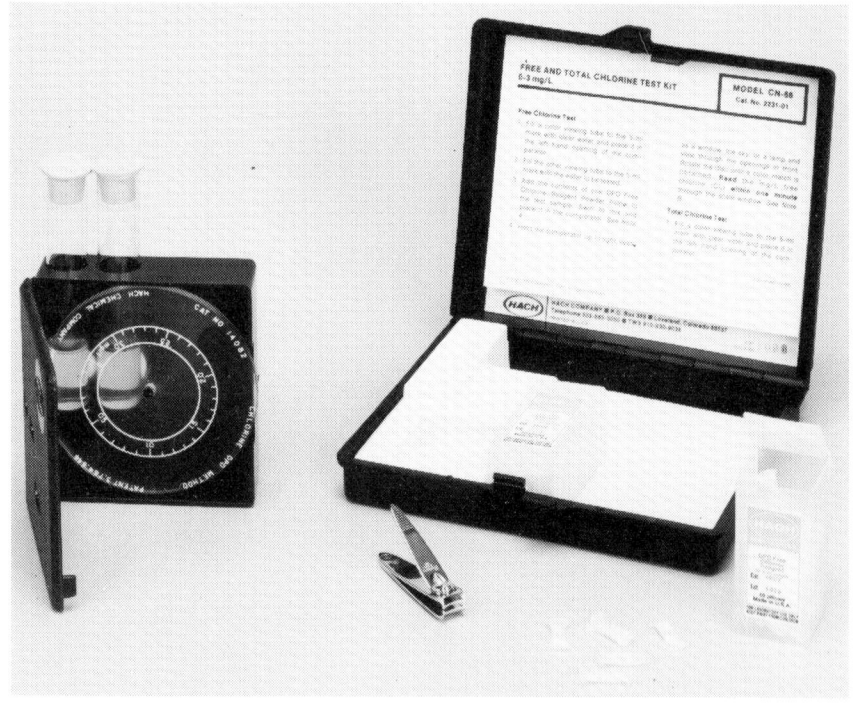

Figure 3.6 HACH Kit CN66 for chlorine testing

Procedure:
Free chlorine test:
1. Fill a colour viewing tube to the 5 ml mark with clear water and place it in the left hand opening of the comparator.

80

2. Fill the other viewing tube to the 5 ml mark with water to be tested.
3. Add the contents of one DPD free chlorine reagent powder pillow in the test sample. Swirl to mix and place it in the comparator.
4. Hold the comparator up to the light, such as a window, the sky, or a lamp and view through the openings in front. Rotate the disc until a colour match is obtained. READ the mg/l free chlorine (Cl_2) WITHIN ONE MINUTE through the scale window centre.

Total chlorine:
As for free chlorine except step 3 above. Add contents of one DPD total chlorine powder pillow to the test sample, swirl to mix and place it in the comparator.

Read off the value AFTER 3 MINUTES and record the value as total chlorine mg/l.

HACH DPD Free Chlorine Test Cube Kit (Not EPA approved):
Kit description: HACH DPD free chlorine test cube kit 20603 – 00. This kit uses the same powder pillow as the previous test but compare the pink colour with the cube colour in 5 stages 0.5, 1.0, 1.5, 2.0, 2.5 mg/l. This is NOT sufficiently accurate for monitoring purposes.
Cost: DPD Test Cube Kit for free Cl_2 with reagent powder pillows for
50 tests: £12.20.

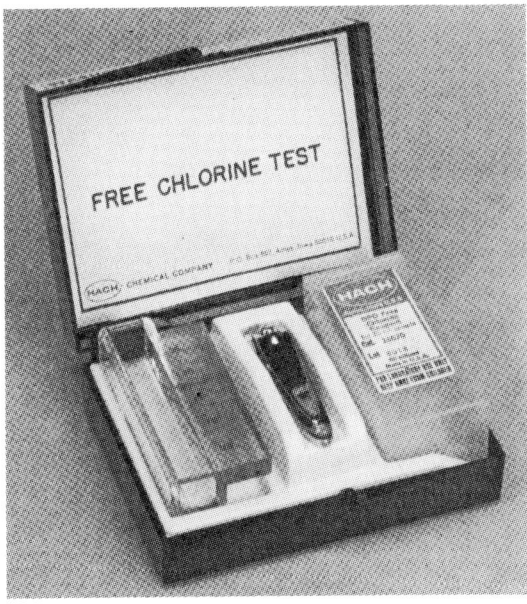

Figure 3.7   HACH cube kit

3.21.2 Chinese starch-potassium iodide method (IDRC, (1981) and Reid (1975))

Equipment required:
Clean glass bottle, 100 ml measuring cylinder, dropping bottle or dropper, starch-potassium iodide solution — 2 grams of soluble starch is dissolved in 100 ml of distilled water, boiled and allowed to cool to room temperature. Eight grams of potassium iodide (KI) is added and the mixture is agitated until the KI is dissolved completely. The solution is best stored in a brown glass bottle in the dark. It should be stable for about 2 weeks if 2 or 3 drops of chloroform (or formaldehyde) are added.

Cost: Cheap.

Procedure: To 100 ml of sample add 5 – 6 drops of the starch/KI solution and mix carefully.

No colour indicates absence of chlorine.

Light blue indicates about 0.1 – 0.3 mg/l Cl_2.

Dark blue indicates above 0.3 mg/l Cl_2.

3.22 Dissolved oxygen

The concentration of dissolved oxygen in waters varies greatly and is dependent on several physical, chemical, biological and microbiological processes. Water in contact with air will contain a quantity of oxygen depending on:

(a) the atmospheric pressure;

(b) temperature of water;

(c) salinity or TDS.

For fresh waters the main variables are temperature and atmospheric pressure. Increased temperature generally lowers the amount of oxygen dissolved in the water and increased atmospheric pressure raises the amount of oxygen dissolved. The solubility of oxygen at various temperatures, pressures and salinities can be found from tables (APHA, 1980), but as an example of the temperature effect on a fresh water at 1 atmosphere, the oxygen content (100% saturated) at 4 °C, 14 °C and 24 °C is 13.11 mg/l, 10.29 mg/l and 8.42 mg/l respectively.

In groundwaters the dissolved oxygen content ranges from zero to 100% saturation. The lower values at depth may be due to oxidation of organic materials depleting the oxygen as the water percolates downwards or may be related to oxidation of iron and manganese. Oxygen deficient groundwaters may need to be aerated (by cascades or weirs) to improve their taste.

In surface waters the dissolved oxygen content is influenced by the degree of biological and biochemical activity. It is one of the most important analyses in determining the quality of rivers and lakes. The suitability of water for fish and other organisms and the progress of self-purification can be measured or estimated from the dissolved oxygen content. The dissolved oxygen content varies with water depth, sludge deposits, temperature, clarity, time and flow regime. The variability of the oxygen content may require a number of samples to be taken to achieve representative values. In developing countries oxygen production by algae may supersaturate the surface layers of lakes and reservoirs during the day so the need for good sampling is paramount.

Dissolved oxygen speeds up the corrosive attack on iron, steel, galvanised iron and brass, especially at acidic pH values.

The determination of dissolved oxygen should be carried out in the field and can even be carried out *in situ* with the latest probes. There are several techniques available for dissolved oxygen determination. The chemical method uses the azide modification of the Winkler iodometric titration method. The instrumental methods involve the use of dissolved oxygen electrodes.

3.22.1 Titration method

(A) Supplier: Camlab Ltd.
Kit description: HACH Dissolved Oxygen Test Kit Model OX – 2P. Cat. No. 1469 – 00. Drop titration method. 1 drop = 1 mg/l as dissolved oxygen.
Sample size: 60 ml.
No. of tests per kit: 100.
Cost: £36.50.

Kit description: HACH Dissolved Oxygen Test Kit with Digital Titrator Model OX – DT. Cat. No. 20631 – 00.
Range: 0 – 10 mg/l as dissolved oxygen.
Smallest increment: 0.01 mg/l.
Sample size: 300 ml.
No. of tests per kit: 50.
Cost: £162.00.

(B) Supplier: Wilkinson and Simpson Ltd.
Kit description: Lamotte Dissolved Oxygen Test Kit Model EDO 7414.
Range: 0.08 – 20 mg/l as dissolved oxygen. Micro burette titration in steps of 0.08 mg/l. Five reagents.
No. of tests per kit: 25.
Cost: £25.20.

(C) Supplier: Lovibond-Tintometer Ltd.
Kit description: Model AF371, Dissolved Oxygen Kit. Uses Lovibond Comparator 2000. 3/3 Colour Disc, 4, 5, 6, 7, 7, 9, 10, 11, 12 mg/l O_2 steps. Wooden case with reagents.
Cost: £97.20.

3.22.2 Dissolved oxygen electrode methods

These electrodes provide an excellent method for continuous, *in-situ* determinations of dissolved oxygen in polluted waters, highly coloured waters and strong waste effluents. Temperature compensation is usually made automatically by thermistors in the electrode circuit. The membranes are best kept moist and need standardising once a day at least. The method of standardisation is simple either using saturated water (by bubbling air through a suitable water) or by exposure to air depending on the instrument. Most electrodes can determine oxygen to the nearest 0.1 mg/l.

(A) Supplier: Camlab Ltd.
Kit description: HACH Dissolved Oxygen Meter 16046 – 02 (240 VAC) with rechargeable Ni/Cd batteries.
Sensing element: Clarke type membrane covered polarographic probe.
Ranges: 0 – 10 and 0 – 20 mg/l oxygen 5 – 45°.
Meter readout, supplied in durable case complete with probe and 3 metres of cable, various spares and instruction manual. Also available with different cable lengths for depth/DO/Temp profiles in lakes etc.
Cost: (with 3 metre cable) £895.00.

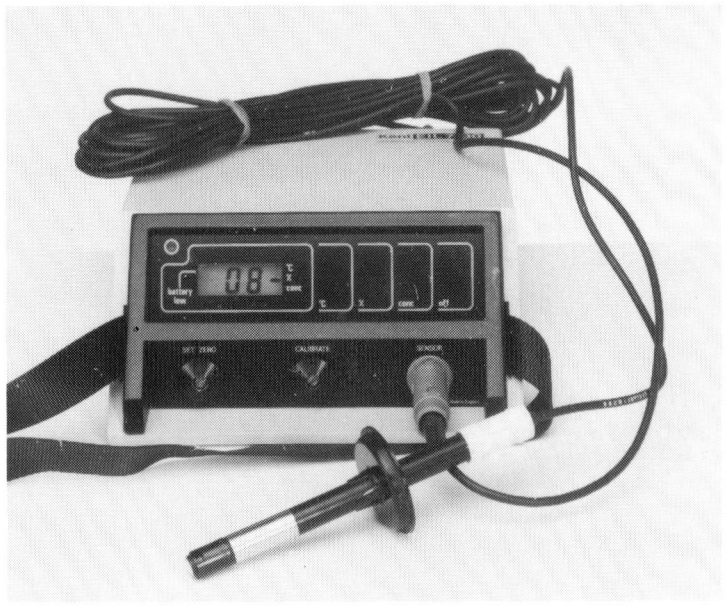

Figure 3.8   Kent dissolved oxygen meter model 7130

Kit description: Schott Gerate O_2 meter CG867.
Digital LCD display for dissolved oxygen and temperature.
Ranges: $0 - 19.9$ mg/l O_2. $0 - 50$ °C temperature.
Automatic temperature compensation for O_2.
The meter is supplied in portable case with electrode, solutions and spare membrane with about 1 metre of cable and instructions.
Dimensions 104 × 50 × 170 mm.
Cost: £299.00.

(B) Supplier and Manufacturer: Kent-EIL Ltd.
Kit description: Kent-EIL Dissolved Oxygen Meter Model 7130. Uses Mackereth sensor with replaceable oxygen capsule and is designed for field and laboratory use since it also fits BOD bottles. Read on LCD Liquid Crystal Display with automatic temperature compensation on mg/l or % saturation. Temperature readout -5 to 50 °C. Touch panel switches for single operation. Rugged streamlined case splash-proof and knock-proof. Batteries and mains operation. Supplied with temperature and oxygen sensor.
Cost: £490.00.
Note: Dissolved Oxygen Electrodes. If an expanded scale pH meter or specific ion meter such as those described in the sections on pH, fluoride and nitrate are available, the purchase of a dissolved oxygen electrode will allow dissolved oxygen measurements to be made easily.

UK Suppliers of dissolved oxygen electrodes:
(A) MSE-Fisons.
Kit description: Orion dissolved oxygen electrode OR 970800.
Cost: £335.00.

(B) Kent-EIL.
Kit description: Dissolved Oxygen probe $8012 - 100$.
Cost: £180.00.

3.22.3 Multiprobe water checker
Supplier: Centronic Sales Ltd.
Instrument description: Horiba Water Checker U-7/2. The Horiba U-7/2 is a hand-held battery-powered instrument designed to check dissolved oxygen, conductivity, temperature, turbidity, and pH. The sensing elements for all five measurements are contained in a single sensor assembly 70 mm in diameter. It can be supplied with various lengths of cable for *in situ* measurements or depth profiles or shallow wells.

The model U-7/2 has an illuminated red digital display and a protective cover over all calibration controls. After calibration using the sensor unit to hold the calibration solution (pH, conductivity and DO) the user simply submerges the sensor (or pours a sample into the chamber) selects the measurement to be made and presses a switch to display the results. An indicator shows if the battery needs recharging.

The specifications are impressive and the author has used this instrument in the field in the UK and Sri Lanka.

Specifications:

Dissolved Oxygen range: 0 – 20 mg/l ± 0.2.

pH 0 – 14 ± 0.1.

Temperature range: 0 – 40 °C ± 0.5 °C.

Turbidity range: 0 – 999 FTU.

Conductivity various ranges 0 – 50000 µs/cm.

This instrument covers the main physical parameters of use in the field. One small difficulty is reading the red digital display in bright sunlight – perhaps the manufacturers will fit a grey LCD to their later models.

Cost: £1200.00.

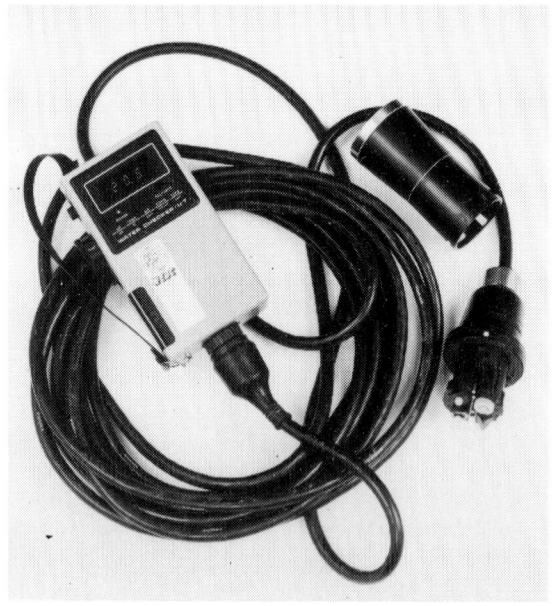

Figure 3.9   Horiba U – 7/2 Water Checker

3.23 Carbon dioxide

Carbon dioxide is dissolved from the atmosphere by rainwater but a larger amount is dissolved by water flowing over or through soil and then entering the groundwater or surface water systems.

In groundwater the concentration of free carbon dioxide may reach 100 mg/l due to biochemical oxidation of organic matter. The equilibrium between carbon dioxide, calcium and bicarbonates at depths where pressure prevents the escape of the carbon dioxide is easily disturbed when pumping begins. The water in the region of the pump, and when it reaches the surface, bubbles with the release of carbon dioxide. This release of carbon dioxide can cause cavitation in pumps and pipework in the well since the water becomes very corrosive (Johnson, UOP 1975). The presence of carbon dioxide in water appears to have no harmful action to humans and in many cases springs with carbon dioxide bubbles are highly regarded as mineral waters and as being good for digestion. The water engineer, however, has to contend with the corrosive properties of free carbon dioxide.

In surface waters the carbon dioxide concentration is unlikely to be greater than 10 mg/l since free carbon dioxide is readily released to the atmosphere. High levels may occur in surface waters where organic degradation due to stagnation or pollution occurs.

Since free carbon dioxide is released quickly on exposure to air, a field determination is most appropriate. Samples need to be collected avoiding contact with air as much as possible.

The determination must be carried out quickly. The sample is best collected by allowing water to run through a measuring cylinder from the bottom up and flushing out the cylinder for several minutes before a rough determination is made. The rubber sampling tube is withdrawn slowly as the water overflows.

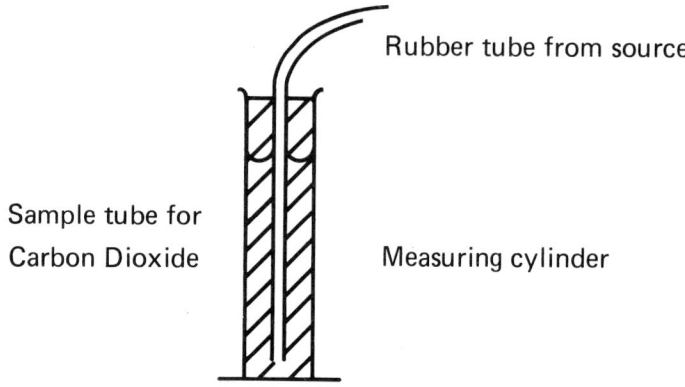

Rubber tube from source

Sample tube for
Carbon Dioxide

Measuring cylinder

Figure 3.10   CO$_2$ sample collection

The determination is carried out by titrating with sodium hydroxide to the pink phenolphthalein end point.

An alternative method is to use the pH-alkalinity relationship provided these are determined in the field with reference to standard graphs and tables.

3.23.1 Titrimetric method

(A) Supplier: Camlab Ltd.

Kit description: HACH Carbon Dioxide Test Kit Model CA – 23 Cat. No. 1436 – 01, titration method, 1 drop = 5 mg/l as CO_2.

Sample size: 5.83 ml.

No. of average tests: 200. In kit.

Cost: £15.55.

Note: This test kit is probably more suitable for low levels of carbon dioxide encountered in surface waters since the small sample size will not give accurate results.

Kit description: HACH Carbon Dioxide Test Kit, Low and High Range with Digital Titrator Model CA – DT. Cat No. 20641 – 00.

Ranges: 0-100 mg/l and 0-1000 mg/l as carbon dioxide.

Smallest increment: 0.1 mg/l and 1 mg/l respectively.

Sample size: 100 ml.

No. of average tests: 100 (50 low range and 50 high range).

Full kit and digital titrator.

Cost: £130.00.

(B) Supplier: Wilkinson and Simpson Ltd.

Kit description: Lamotte Model PCO – DC Carbon Dioxide Test Kit Code No. 7525.

Range: 0-250 mg/l as carbon dioxide. Each drop equals 2.5 mg/l. Phenolphthalein end point. Test kit for 25 tests.

Cost: £10.00.

3.23.2 Calculation method

Equipment required:

pH equipment.

Total alkalinity equipment (see relevant sections).

Procedure A:

The nomogram below for US Air Force (1959) is sufficient for most purposes but for more accurate values calculations are outlined from APHA (1980).

Use of Figure 3.11:

1. Locate the *diagonal* line for the known pH value.

2. Follow along this line to the point where it crosses the *vertical* line for the known alkalinity.

88

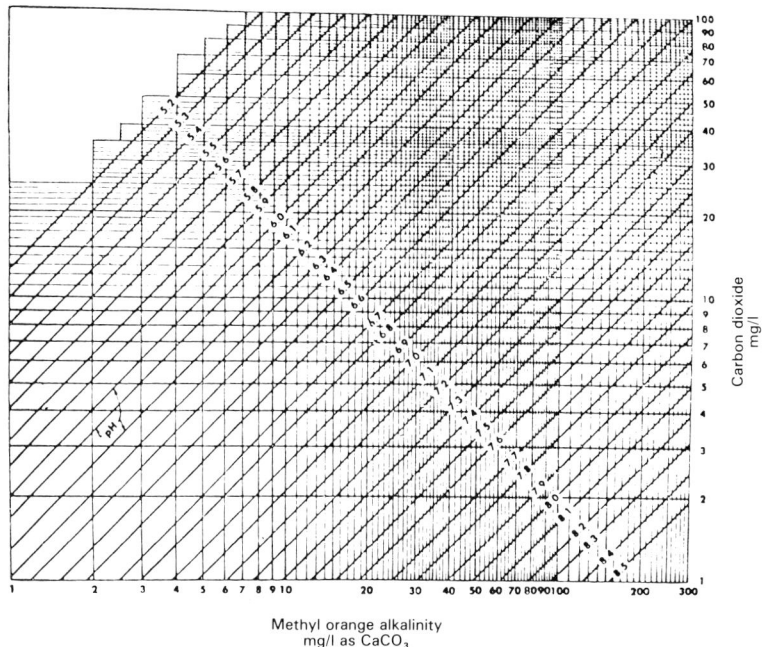

Carbon dioxide mg/l

Methyl orange alkalinity
mg/l as CaCO₃

Figure 3.11

3. The *horizontal* line passing through this point of intersection gives the content of carbon dioxide in parts per million.
Example: Assume the pH to be 6.6 and the alkalinity to be 19 mg/l. The horizontal line passing through the point of intersection of the lines for these two values show, on the scale at the right, a carbon dioxide content of 10 mg/l.

Procedure B:
APHA (1980) computational method for waters with TDS below 500 mg/l at 25 °C.
Compute the forms of alkalinity and sample pH and total alkalinity using the following equations:
(a)   Bicarbonate akalinity:
$$HCO_2^- \text{ as mg } CaCO_3/l = \frac{T - 5.0 \times 10^{(pH-10)}}{1 + 0.94 \times 10^{(pH-10)}}$$
where: T = total alkalinity mg $CaCO_3/l$.
(b)   Carbonate alkalinity:
$$CO_3^{2-} \text{ as mg } CaCO_3/l = 0.94 \times B \times 10^{(pH-10)}$$
where: B = bicarbonate alkalinity, from (a).

89

(c) Hydroxide alkalinity:

OH^- as mg $CaCO_3/l = 5.0 \times 10^{(pH-10)}$.

(d) Free carbon dioxide:

mg $CO_2/l = 2.0 \times B \times 10^{(6-pH)}$

where: B = bicarbonate alkalinity, from (a).

(e) Total carbon dioxide:

mg total $CO_2/l = A\ 0.44\ (2b + C)$

where: A = mg free CO_2/l

B = bicarbonate alkalinity from (a), and

C = carbonate alkalinity from (b).

4. Physical Analysis

4.1 pH

pH is used to describe the *intensity* of acidity of a solution and is defined as 'the logarithm to base 10 of the reciprocal of the hydrogen ion activity expressed in moles per litre'.

$$pH = \log_{10} \frac{1}{[H+]}$$

For water this results in a scale from 0-14 units with neutral solutions at pH 7 and acidic solutions from pH 0-7 and alkaline or basic solutions from pH 7-14. pH varies with temperature. The WHO Guide level for pH is 6.5 to 8.5 (WHO, 1983).

Most natural waters have a pH from 4 to 9 and the majority are slightly alkaline (above 7) due to carbonates and bicarbonates of calcium and magnesium dissolved in the water. Waters with a pH outside the range 6-9 are likely to be contaminated indicating the intrusion of strongly acidic or alkaline wastes (probably industrial).

Many water supply treatment processes are pH dependent such as coagulation, disinfection, softening and corrosion control. The measurement of pH is one of the most frequently used tests in water chemistry. It is also important to perform pH tests as soon as possible after sampling since biological and chemical reactions between the atmosphere and the sample can rapidly alter the pH. Reactions such as iron deposition or carbon dioxide release are two such reactions. So field determination of pH is even more relevant than laboratory determinations.

There are three basic measurement techniques which can be used in the field, depending on the accuracy required and money available. These are:

(a) pH indicator papers—
Wide range 0-14, and narrow range, and 2 or 3 pH units.

(b) Colour comparators—wide range, narrow range.
Discs and cards.

(c) Portable pH electrodes and meters.
Dipstick
Conventional
Expanded scale

4.1.1 pH indicator papers

There are several manufacturers of pH indicator papers throughout the world, according to the International Laboratory 1981 Buyers Guide, the most notable *UK suppliers* being:

BDH Chemicals Ltd.
Kit description: Whatman-BDH. Merckoquant.

Whatman Ltd.
Kit description: Whatman Test papers.

Camlab Ltd.
Kit description: Macherey-Nagel.

These suppliers should be contacted directly for more specific details of their test paper range. It is suggested that non-bleeding test strips are used if possible.

Cost: 1 box of 100 pH test strips costs approximately £2.20.
They should be kept stored away from sunlight and should last for about 2 years.
Equipment required:
1 box of wide range pH paper.
Boxes of narrow range papers cover ranges pH4 – pH9.5. The graduations of pH in the narrow range papers are about 0.2 or 0.3 pH units.

4.1.2 Colorimetric indicator methods

The action of various dyes, called indicators, which change colour in a defined, reproducible way with the pH of the solution in which they are mixed is the basis of colorimetric pH measurement. The most simple is the reaction of litmus, red for acid and blue for alkaline solutions. Various dyes and mixtures of dyes are used to cover pH ranges from 1.2 to 14. The indicators used for natural waters are usually Universal (a mixture covering pH 3-11), phenol red (pH 6.5-8.5), bromothymol blue (pH 5.5-8.5), thymol blue (pH 7.8-10), methyl red (pH 4.4-6). They are supplied either as solutions in dropper bottles or tablets in aluminium foil packs.

The colours can be compared against either printed cards (increments ± 0.5 pH) or colour discs (increments ± 0.1 or 0.2 pH units) or colour cubes (increments 0.5 pH units).

Note: Chlorine in amounts greater than 1 mg/l interferes with pH indicators and is best removed by using one drop of 0.5% sodium thiosulphate ($Na_2S_2O_3$) solution, prior to testing.

Equipment required: Printed comparator cards.

There are several inexpensive pH indicator and colour comparison cards available in the UK.

Supplier: Wilkinson-Simpson Ltd.
Description: CS 129 for pH 4-11. Universal pH kit (100 tests).
Cost: £4.00.

Figure 4.1   HACH wide range pH test kit

Supplier: Camlab Ltd.
Description: UNISOL Range. Unisol 410 pH range 4-10. Cat. No. 91002.
Cost: £5.00.

Supplier: BDH Chemicals Ltd.
Description: Aquamerck 8038, pH range 4.5-9.
Cost: £12.00 (includes 5 plastic measuring cells).
Equipment required:
There are two major manufacturers of colour comparators: Lovibond-Tintometer Ltd. and HACH Chemicals Co. The Lovibond equipment incorporates colour discs with permanent non-fading glass standards and the pH indicator is usually in tablet form wrapped individually in aluminium foil packets. HACH pH standard colour discs are plastic with a continual colour change and the pH indicator is usually supplied in plastic dropper bottles, sufficient for about 100 tests.

Lovibond pH tests: *Cost:* £22.80
Lovibond 2000 Comparator.

Disc. no.	pH range	Indicator	
2/1P	4 – 11	Universal	
2/1E	4.4 – 6.0	Methyl Red	
2/1H	6.0 – 7.6	Bromothymol Blue	£19.00
2/1J	6.8 – 8.4	Phenol Red	
2/1L	8.0 – 9.6	Thymol Blue	

For other discs refer to Lovibond Ltd.
Cost: of 100 indicator tablets in foil strip pack: £1.90.

HACH '17' Series Test Kits (includes reagents, discs and colourmeters).

Cat. No.	pH range	Indicator	Cost
1470 – 11,17N	4 – 10	Wide range pH	£42.25
1470 – 06,17F	5.5 – 8.5	Bromo Thymol Blue	£32.50
1470 – 08,17H	6.5 – 8.5	Phenol Red	£32.50
1470 – 09,17J	7.8 – 10.0	Thymol Blue	£32.50

For other discs refer to HACH or Camlab (UK Agents).
Note: For unknown samples it is useful to use the Universal indicator to give an idea of the pH range to be used for the more accurate determination. pH paper strips may be sufficiently accurate for the initial screening.

4.1.3 Field pH meters
pH can be determined electronically more accurately and with much less interference than by using colorimetric methods. pH is measured

by comparing and amplifying very small voltages or potentials produced by immersing a sensing (or glass) electrode and a reference electrode in a solution. The glass electrode consists of a special sensitive glass bulb containing a fixed concentration of hydrochloric acid solution in contact with an internal reference electrode producing small voltages in response to pH changes (about 59 mV per pH unit).

The glass electrode and reference electrode are often combined for easier handling, strength and field work applications.

The increasing use of solid state circuitry, liquid crystal displays and microelectronics has miniaturised pH instrumentation so much that hand-held pH meters are available which can read ±0.1 pH. For greater accuracy the amplification circuits and meters need to be larger and expanded scale field pH meters are available to read ±0.02 pH unit (a response to 1 mV). The expanded scale meters are more useful for determining parameters such as fluoride and nitrates in the field since pH accuracy of ±0.02 pH units is not usually needed in most water treatment processes.

There are several manufacturers of field pH meters and the costs vary according to size and accuracy required. pH meters are often combined with multipurpose instruments to determine other parameters such as conductivity, temperature, etc.

Electrometric pH measurements are subject to temperature effects and require to be standardised regularly using pH buffer solutions. Buffer solutions at pH 4, 7 and 9 can be easily prepared by dissolving powder or tablets in a volume of deionised water. The pH meter is then buffered by immersing the electrodes into the buffer solution and adjusting the scale reading or LCD reading by turning the buffer knob or standardise control.

For field work it is best to avoid red crystal dislays as they cannot be seen in strong sunlight. Use a rugged combination electrode and store the pH meter in a travelling box or case after use. Always carry a spare battery (or ensure regular recharging), spare electrodes and pH buffer tablets. For most waters pH buffer tablets will be most useful. They can be supplied by BDH Chemicals Ltd., Cat. No. 33155, 25 Buffer tablets pH 7.0 at a cost of £3.50 per 50 tablets. There are 3 types of field pH meters: the dip probe, the conventional type of meter and the expanded scale field pH meter.

4.1.3.1 Dip probe meters
Some examples currently available in UK:
Supplier: Camlab Ltd.
Kit description: Camlab hand-held digital pH meter with Ni-Cd rechargeable battery (recharger supplied). SA/LD014.
Cost: £165.00.

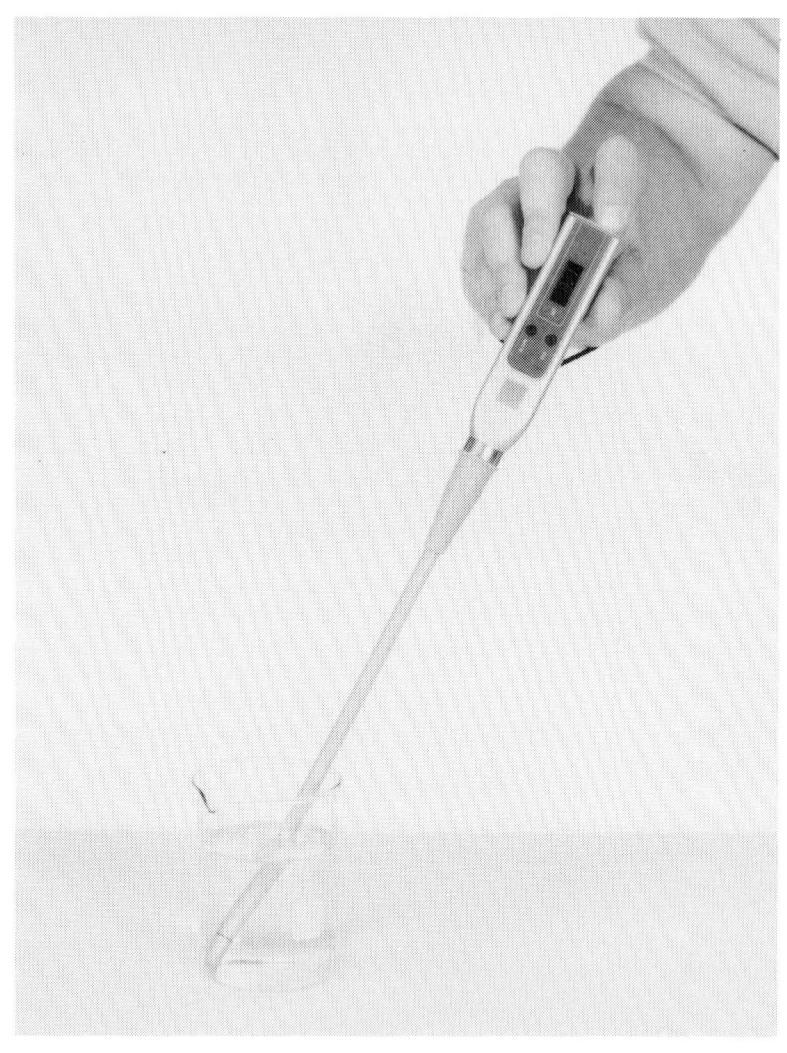

Figure 4.2   Camlab pH stick

Supplier: A. Gallenkamp Ltd.
Kit description: Gallenkamp 'pH stick', automatic temperature compensator, splash proof. Protection wallet, 275 × 35 × 30 mm dimensions, 100 g wt.
7 × 1.4 volt mercury cell (hearing aid batteries).
Range: 0-14 pH ±0.03 pH.
pH 4 and 7 buffers supplied.
Cost: £125.00.

4.1.3.2 Conventional field pH meters
Supplier: Camlab Ltd.
Kit description: HACH mini pH meter. Cat. No. 17200 – 10.
Cost: £180.00.

Supplier: MSE-Fisons Ltd., MSE Scientific Instruments.
Kit description: Orion Model 201 portable digital pH meter complete with carrycase, gel-filled unbreakable combination electrode pH 7 buffer, 220 volt AC line converter, instruction manual.
Cost: £180.00.

Supplier: pHOX Systems Ltd.
Kit description: pHOX Model 42E pH meter.
Range: 0-14 pH ± 0.01 pH.
Environmentally protected case (200 × 125 × 275 mm) with external switches. Recorder output. Battery powered. Includes pH probe, analogue readout.
Cost: £150.00.
pHOX Model 67 multiprobe meter for pH, D.O., temperature, conductivity, redox potential. Battery powered, environmentally protected with 2 channel output, analogue readout.
Cost: £700.00.
pHOX Systems Ltd. also make customer specified equipment.

4.1.3.3 Expanded scale field pH meters
Two varieties known to be available in UK:
Supplier: Camlab.
Kit description: HACH expanded range pH meter and electrodes and carrycase. Cat. No. 16400 – 02.
Cost: £550.00.

Supplier: MSE-Fisons Ltd, MSE Scientific Instruments.
Kit description: Orion 407 A/F specific ion meter, pH range 0-14, in carrying case with rechargeable battery pack *but no electrodes.*
Cost: £554.00, pH electrodes cost an additional £80.
Procedure:
Refer to maker's instruction manual for details of operation and calibration of pH meters.

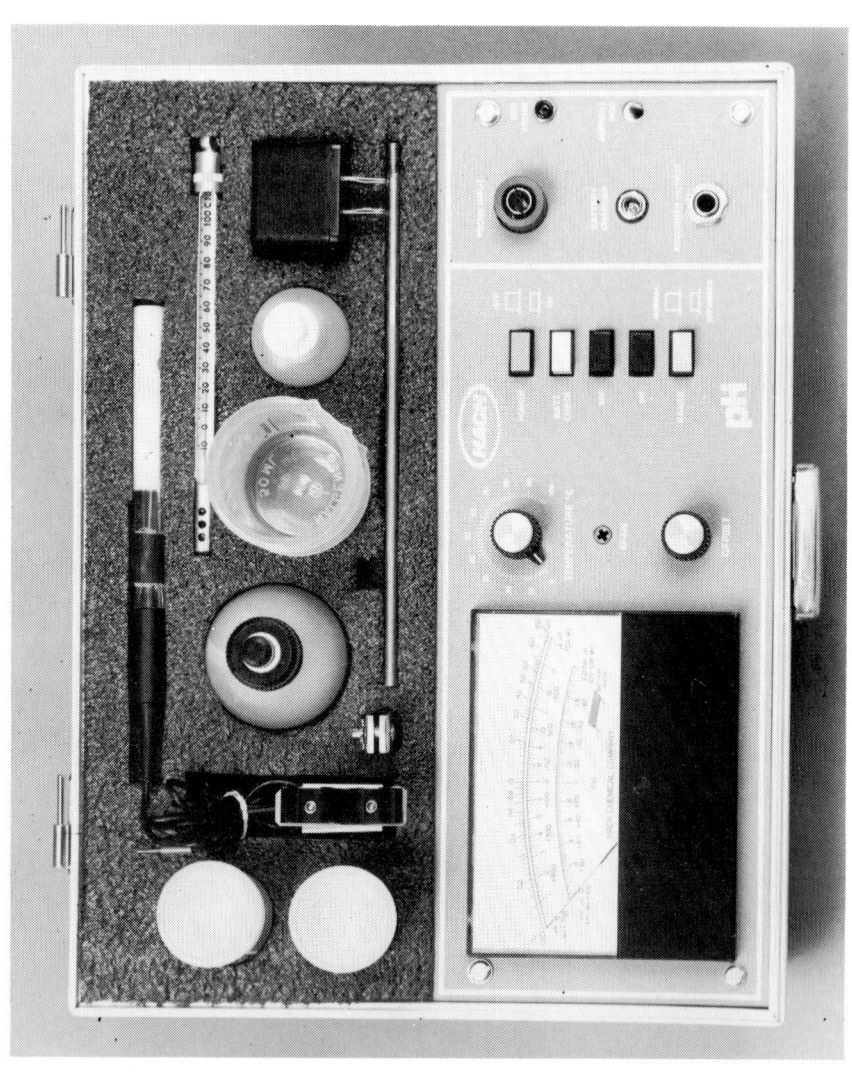

Figure 4.3   HACH expanded scale pH

98

The field pH meters should be buffered at least once a day when in use.

See also Section 3.2.2, Dissolved Oxygen, for details of Horiba Water Checker for pH measurements.

It is important to ensure good connections between the pH meter and electrodes. Many faults can be ascribed to loose electrode connections especially in field conditions in hot climates. There are a variety of connections possible and it is good to check their soundness by touching them and observing the meter readings.

4.2 Redox potential, Eh

The redox potential is used particularly in respect of groundwater systems as an aid to predicting corrosion due to soluble iron. It is carried out on a sample avoiding exposure to air and usually a flow-through cell arrangement is used as described below.

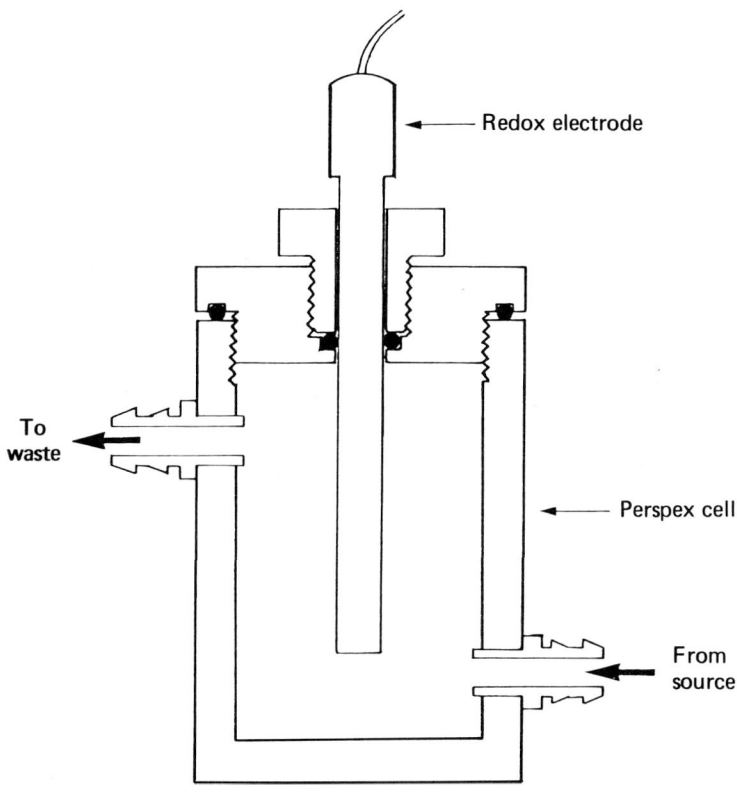

Figure 4.4 Flow-through cell for Eh

The method is that described by Cook and Miles (1980) and requires reagents to be prepared by a support laboratory.

4.2.1 Equipment required

mV/pH meter: the expanded scale pH meters described in the pH section are suitable.
Supplier: Camlab Ltd.
Kit description: HACH expanded scale field pH meter. Cat. No. 16400 02.
Cost: £550.00.

Supplier: MSE-Fisons Ltd.
Kit description: Orion 407 A/F portable specific ion meter.
Cost: £554.00.

Platinum redox combination electrode:
Supplier: MSE-Fisons Ltd.
Kit description: Orion 96-78.
Cost: £60.00.

Flow through cell
Flexible tubing and connections.
Thermometer (0-100 °C).

4.2.2 Reagents

Reference Solution A, 0.1M potassium ferrocyanide 0.05M potassium ferricyanide.
Dissolve 4.22 g of potassium ferrocyanide ($K_4Fe(CH)_6.3H_2O$) and 1.65 g of potassium ferricyanide ($K_3Fe(CN)_6$) in deionised water and dilute to 100 ml. Stored in a dark bottle, this solution is stable for several weeks and should be discarded as soon as any growth or precipitate appears.

Reference solution B, 0.01M potassium ferrocyanide 0.05M potassium ferricyanide 0.36M potassium fluoride.
Dissolve 0.42 g of potassium ferrocyanide ($K_4Fe(CN)_6.3H_2O$) 1.65 g of potassium ferricyanide ($K_3Fe(CN)_6$) and 3.39 g of potassium fluoride ($KF.3H_2O$) in deionised water and dilute to 100 ml. Stored in a dark bottle, this solution is stable for several weeks, but discard if any precipitate or growth appears.

Zobell's solution
Dissolve 1.408 g of potassium ferrocyanide ($K_4Fe(CN)_6.3H_2O$) and 1.65 g of potassium ferricyanide ($K_3Fe(CN)_6$) and 7.46 g of KCl in

100

deionised water and dilute to 1 litre. Stored in a dark bottle, this solution is stable for several weeks but discard if any precipitate or growth appears.

Electrode filling solution
For measurements in dilute groundwaters of below 0.2M total ionic strength, fill the electrode with Orion solution 90 – 00 – 01. For waters above 0.2M total ionic strength use Orion solution 90 – 00 – 11 (4M KCl saturated with AgCl).

Procedure:
1. Fill the electrode and connect to the meter according to the manufacturer's instructions.
2. Check the electrode using either Zobell's solution or reference solutions A and B. The potential developed in Zobell's solution should be close to 210 mV at 10 °C and 190 mV at 20 °C. In reference solution A, a reading of 192 mV should be obtained and reference solution B should give a reading approximately 66 mV greater.
3. Rinse the electrode and insert it in the cell, making sure that the seal is airtight.
4. Connect the cell to the wellhead by means of the flexible tubing, ensuring that all connections are airtight.
5. Allow water to flow through the cell at a constant rate of about 1.5-2 l/min. and tilt the cell to expel any trapped air bubbles. Note the temperature of the water.
6. Record the mV readings at 1 minute intervals until a steady reading is obtained, typically within 20-30 minutes. If the meter has an expanded relative mV scale, this may be used with advantage to monitor the drift in reading.
7. Switch back to the absolute mV scale. Stop the water flow to remove any streaming potential and immediately record the mV reading.
8. Calculate the Eh value from: Eh = absolute mV + reference electrode potential.
The potentials developed by the reference electrode of the Orion 96-78 electrode when filled with 90 – 00 – 01 solution at various temperatures are:

Temperature in °C	Reference electrode potential mV
0	+257
10	+251
20	+244
25	+241
30	+238

4.3 Conductivity

A rapid method of estimating the dissolved salts in a water sample is by measurement of its electrical conductivity (Holden, 1970). The conductivity is related to the total concentration of ions in solution, their valency (charge), and mobility and to the temperature of measurement. Conductivity is more related to inorganic components of a water sample than less dissociated organic components.

Changes in conductivity of a sample may signal changes in mineral composition of raw water, seasonal variations in reservoirs, intrusion of sea or saline waters by overpumping, and pollution from industrial wastes.

Since ions have different mobilities the relationship between total dissolved solids and conductivity is not constant for all waters but varies in the range below:

$$\text{Conductivity} \times \text{Factor (0.55 to 0.9)} = \text{TDS, mg/l}$$

(expressed μS/cm, or microohms/cm)

Total Dissolved Solids

(Equation 1)

The units of conductivity used to be expressed as micromho per centimetre (μohms/cm) but in SI units the reciprocal of ohms is expressed as Siemens and the unit for conductivity should be expressed as millisiemens per metre (mS/m). (APHA, 1980).

$$1 \text{ mS/m} = 10 \text{ mhos/cm}$$

(Equation 2)

This is confusing so the units are best expressed as microsiemens per centimetre according to HMSO (1979) recommendations, since the actual conductivities are small and the distance between electrodes is usually also 1 cm.

$$1 \text{ μS/cm} = 1 \text{ μmho/cm}$$

(Equation 3)

Conductivity readings are also expressed at different temperatures: 25 °C in USA and 20 °C in UK. The variation of conductivity with temperature is an increase of 1.9 per °C.

It is important to measure the temperature of the sample while measuring its conductivity. Several instruments automatically compensate for temperature.

Different instruments have different scales, cell constants and performance characteristics and it is useful to calibrate them, especially if several instruments are being used in one project so their readings can be related. It is useful to calibrate them against potassium chloride standard at 20 °C or 25 °C. The concentration of standard KCl used is in the range 0-1 Normal but for ready reference a 0.01

Normal solution of KCl (0.7456 g/litre) at 25 °C has a conductivity of 1412 µS/cm or 1412 µmho/cm and 1279 µS/cm or 1279 µmho/cm at 20 °C.

Some electrodes (metal) require replating with platinum black as this is awkward in the field they should be avoided.

Oils, fats and greases can coat electrodes, therefore electrodes should be cleaned regularly with detergent and water.

4.3.1 Equipment required

1. Accurate thermometer 0-50 °C ± 0.1 °C.
2. Conductivity meter and electrodes.

There are many manufacturers of conductivity meters. Some instruments are cheap and often a little fragile for field use so its worth paying a little extra for a good model. The list below describes some easily available in UK.

Supplier: Camlab Ltd.

Instrument description: Myron L meters. Various ranges but probably most useful 0-5000 µmho/cm.

Automatic temperature compensation (to 25 °C). Battery powered. Hand held, self contained.

Cost: £90.00.

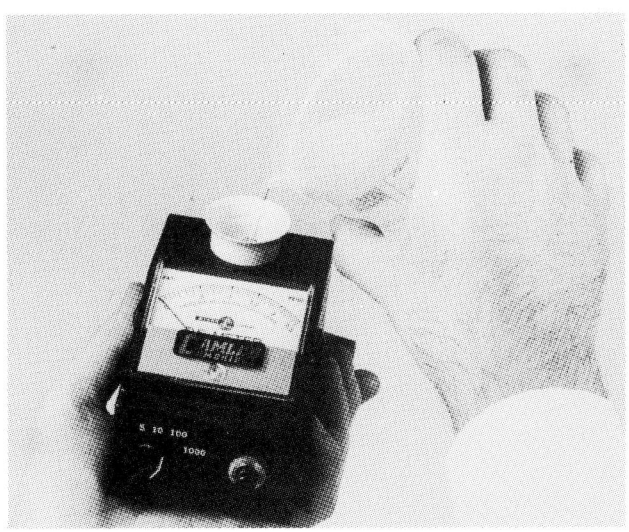

Figure 4.5   Myron conductivity meter

103

Instrument description: HACH mini conductivity meter, 3 ranges 0-100, 0-1000, 0-10000 µmho/cm. Manual temperature compensation. Thermometer battery powered (6AA cells). In plastic carry case. Dip probe.
Cost: £199.00.

Supplier: Wilkinson and Simpson.
Instrument description: Lamotte model DA – D5. Dip probe, 9 volt DC battery with check circuit, carry case.
Range: 0 – 5000 µmho/cm. Manual temperature compensation.
Cost: £198.00.

Supplier: Data Scientific.
Instrument description: PT 1 – 7 portable analogue conductivity meter with cell. Not robust. Needs good protection for field use.
Cost: £65.00.

Supplier: Kent Industrial Measurements Ltd.
Instrument description: Model 5013, lightweight portable conductivity meter. Battery powered. Robust and compact. Supplied with thermometer, case and beaker cell. 0.1 to 10000 µS/cm. Null meter read out.
Cost: £145.00.

Instrument description: Model 5003, portable conductivity meter, battery powered, manual temperature compensation, 2 beaker cells. Range: 0.01 – 100 000 µS/cm. Compact, shockproof, wooden case.
Cost: £225.00.
Note: See also Section 3.22 Dissolved Oxygen for details of Horiba Water Checker for conductivity measurements.

4.4 Temperature

Water temperature is often a good indicator of contamination. Any sudden change in temperature of groundwater suggests that the water is contaminated, possibly from industrial discharges. Temperature has a marked effect on bacterial and chemical reaction rates in water and can control the initiation of algal blooms, the degree of dissolved oxygen saturation and carbon dioxide concentrations. The temperature of water from a household tap may give useful information as tastes may arise from warming of the water due to heating from hot water pipes or exposed tanks. Differences in temperature, between the first running and that after several litres have been drawn, can be recorded.

Mercury filled centrigrade thermometers range 0 – 100 °C will suffice for most procedures.

For borehole temperature measurements *in situ,* or depth/temperature profiles in lakes or reservoirs, electrical resistance

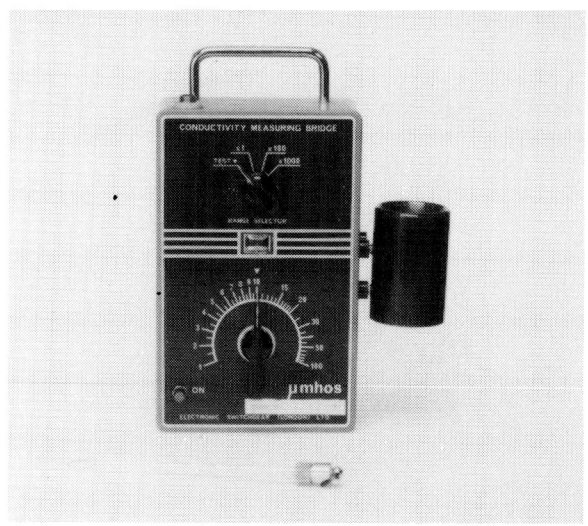

Figure 4.6 EIL conductivity meter model 5013

thermometers with cable and readout may be required. Refer to Section 3.22 Dissolved Oxygen for details of *in-situ* temperature measurements.

Safety Note: Mercury thermometers are fragile and the mercury from a broken thermometer should not be allowed to enter the distribution system.

There are some thermometers available with brass protection and a brass weight which acts as a heat store while the thermometer is being read.

4.4.1 Equipment required
Supplier: (a) Wilkinson and Simpson Ltd.
Kit description: Thermometers in brass protecting case.

Cost: PT 684 range – 5 °C to 50 °C	£6.45
PT 682 range – 10 °C to 110 °C	£5.90

Replacement thermometers (glass):

PT 683 range – 5 °C to 50 °C	£4.35
PT 681 range – 10 °C to 110 °C	£3.75

Supplier: (b) Camlab Ltd.
Kit description: Thermometer 305 mm long.

Cost: THR 110L – 10 °C – 119 °C	£2.85
THR 305B Brass case for above	£6.50

4.5 Turbidity

Turbidity of natural waters is caused by the presence of components such as clay, mud, organic material, bacteria, algae, lime, or rust held in colloidal suspension. It is measured by determining the amount of light scattered and absorbed by the suspended material, usually at 90 °C (right angles) to a source of light.

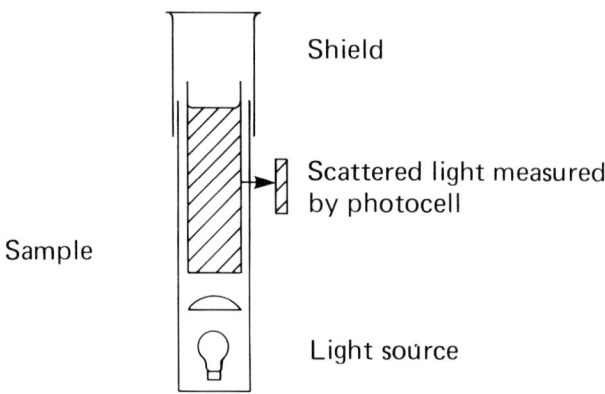

Figure 4.7   Principle of turbidity measurement

Several standards and measuring devices have been used over the years. The first standards were silica from diatomaceous earth, then Fuller's earth (a clay) but both suffered from large errors in reproduction of consistent standards. With the material now used, Formazin, turbidity standards can be prepared repeatedly with an accuracy of ± 1 per cent.

HACH (1979) points out that there is a linear relationship between the turbidity value and the concentration of suspended matter in mg/l, perhaps due to the changing and changeable nature of turbid particles. Secondly he points out that different types of turbidimeter, even when standardised with Formazin 'will give far different turbidity readings when used to measure turbidity of a sample containing another turbidity substance. Variations under the circumstances can be as much as 500%'.

There are no very simple, cheap methods of determining turbidity accurately in the field apart from perhaps visual estimation of a sample with standard prepared Formazin solution such as 10 NTU and 100 NTU.

In developing countries the visual appearance of a drinking water may be more important to the consumer than a definitive turbidity measurement. Turbidity measurements may be of use to overseas water treatment engineers for checking the efficient operation of filters and sedimentation tanks.

Drinking waters should have low turbidities where possible because suspended particulate matter provides suitable sites for the growth of bacteria and other micro-organisms.

4.5.1 Equipment required

At present the HACH 16800 portable turbidimeter is the only portable turbidity meter produced that is known to the author. It is easy to use and is suitable for use in the laboratory and field. It is powered either by its own 6V rechargeable battery or from 110V AC or 240V AC mains supply, and is supplied with a 10 NTU standard. It is suggested that spare standards and sample tubes be purchased since unless well cared for they will become scratched and require renewal.

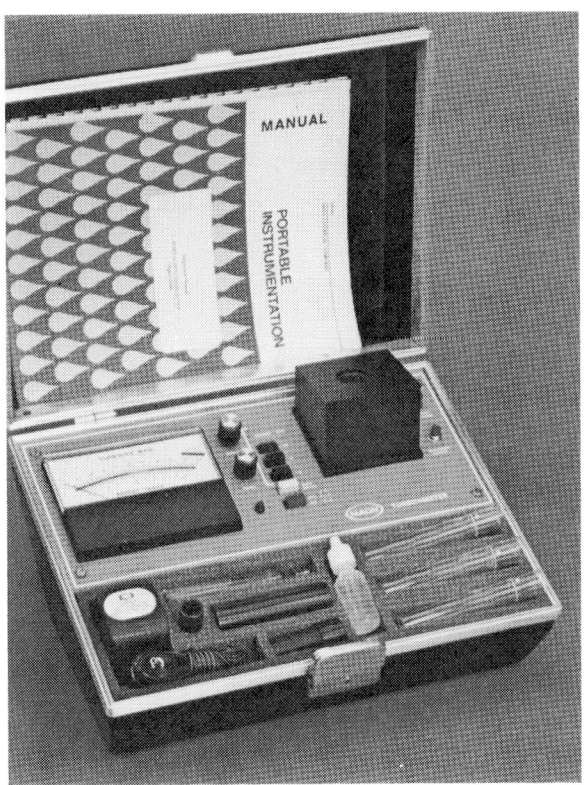

Figure 4.8   HACH 16800 turbidity meter

107

UK Supplier: Camlab Ltd.
Instrument description: HACH Portalab Turbidimeter 16800−02 (240V AC).
Cost: £770.00.

Turbidity as determined	Report to nearest
1.0 or less	0.1
1 − 10	1
10 − 100	5
100 − 400	10
400 − 700	50
over 700	

Note: See also Section 3.22 Dissolved Oxygen for details of Horiba Water Checker for turbidity measurement.

4.6 Tastes and odours

The senses of taste and smell are related. Unpolluted water has no taste or smell so odours and tastes in water indicate contamination.

People have different sensitivities to tastes and odours; some people can taste water containing only 170 mg/l of sodium chloride, others would scarcely notice it at 1400 mg/l (Holden, 1970). Some people become even more sensitive when they know a certain substance is present, for example salt, chlorine or iron but fail to detect it without that knowledge.

The water temperature is also important when testing for tastes and odours. Cool waters are more pleasant to drink than warmer samples. If sea water is chilled to 0 − 2 °C, taste mechanisms are desensitised and the salinity cannot be detected. The temperature of the sample being tested for odour and taste therefore must be recorded.

It is possible for waters to be odourless but possess taste as shown in Table 4.1 but it is best to assess these parameters together to judge if the water is polluted.

4.6.1 Taste

WATER SAMPLES SHOULD NOT BE TASTED UNLESS THEY HAVE BEEN PROVED SAFE FOR DRINKING BY BACTERIOLOGICAL AND CHEMICAL TESTS.

It is often possible in remote areas in developing countries to test the water by drinking tea. Tea is sensitive to salinity and will turn black and become bitter in very saline water and also if iron or manganese is present in the source water. Hopefully the water will be boiled sufficiently for tea preparation to kill all the pathogenic bacteria, however, this may not guarantee the destruction of cysts. The use of coffee for testing salinity is not recommended as the taste of coffee is often improved by adding salt.

108

Table 4.1   Tastes and their possible sources

Taste	Possible source of taste
Salty	High chlorides and sodium salts
Bitter	Magnesium sulphate
Astringent	Calcium sulphate and hard waters iron and manganese
Sweet	Organic matter
Metallic	Iron, manganese, copper, zinc
Sharpness (pleasant)	Chalk derived waters, calcium carbonate and carbon dioxide
Flat, insipid	Surface sources, deficient in oxygen, long stagnation in dead end mains or cisterns
Musty	Fungi formed because of water being heated in transmission system, especially tall buildings, offices

4.6.2   Odours

Water polluted by excreta emits characteristic unpleasant odours. Table 4.2 describes various odours in water (AWWA, 1978).

Table 4.2   Qualitative descriptions of odours

Nature of odour	Description (such as odours of)
Aromatic (spicy)	camphor, cloves, lavender, lemon
cucumber	*Synura*
Balsamic (flowery)	geranium, violet, vanilla
geranium	*Asterionella*
nasturtium	*Aphanizomenon*
sweetish	*Ceolosphaerium*
violet	*Mallomonas*
Chemical	industrial wastes or treatment chemicals
chlorinous	free chlorine
hydrocarbon	oil refinery wastes
medicinal	phenol and iodoform
sulfuretted	hydrogen sulfide or rotten eggs
Disagreeable	(pronouncedly unpleasant)
fishy	*Uroglenopsis, Dinobrvon*
pigpen	*Anabaena*
septic	stale sewage
Earthy	damp earth
peaty	peat
Grassy	crushed grass
Musty	decomposing straw
mouldy	damp cellar
Vegetable	root vegetables

The sources of these odours can be due to:

1. Presence of hydrogen sulphide in water due to septic sewage or chemical and biological reduction of sulphate, especially in deep groundwaters or poor distribution systems.
2. Contamination by organic matter from sewage, septic tanks, pit toilets, soil and vegetation.
3. Contamination by chemicals from industrial sources or treatment processes.
4. Growth of algae, protozoa and fungi in the water.
5. Contact with painted surfaces, such as bituminous linings of mains and tanks.

Taste and odours are often brought to the notice of water authorities by consumers. The cause should be investigated quickly as conditions may rapidly deteriorate if delays occur. The complaints are frequently useful to the authority in detecting pollution events which otherwise might go undetected for months. Often the contamination is within the consumer's plumbing system, but all complaints should be treated seriously.

Equipment required for estimation of odour:
Clean glass bottles, 500 ml wide mouth with glass stoppers, thermometer, water heating device.

Procedure:
1. Take a representative sample of water and half fill the glass bottle. Replace the stopper.
2. Vigorously shake the bottle for 10 – 20 seconds.
3. Remove the stopper and sniff the odour. Record the odour as described in Table 4.2, and note the temperature of the sample.
4. To confirm the odour replace the stopper and heat the sample in a water bath (or a pan of hot water) to 60 °C. Shake the bottle wrapped in a cloth (it will be hot) and again smell the contents.

The intensity of the odour is reported as follows:

Odour intensity

1. odourless   4. obvious
2. slight   5. strong
3. mild   6. very strong

(Institute of Health, China, 1981).

According to Chinese practice in Rural Water supply (IDRC, 1981), if the water emits odours, the source should be investigated before being used further.

The test is by no means quantitative or accurate. It is subject to individual sensitivity and to overcome this the majority opinions of several testers can be sought. More details of sophisticated tests for odour and taste may be found in Standard Methods (AWWA, 1980).

However, it is a cheap test that can be done in the field and is a useful indicator of possible pollution.

4.7 Colour

Colour in water may be due to several forms of pollution. Normally the colours produced are yellows and browns and may be due to:
1. Decaying organic matter of vegetable or soil origin. The colour is very much pH dependent.
2. Colloidal and soluble iron and manganese.
3. Chromate wastes which give only yellow colours.

True colour is due to dissolved or colloidal material substances but suspended material can give an apparent colour. The method of measurement determines the true colour only and should only be applied to clear waters, perhaps after settling.

The method uses matching glass or plastic standards derived from colours produced by solutions of chloroplatinate/cobaltous chloride standards. The samples are viewed in long glass tubes to intensify any colour present which is measured in Hazen units.

Colour is especially important as an aesthetic criterion for drinking water, in addition measurements greater than 5 Hazen units are objectionable in water used for laundries, dairies and paper, textile and food processing industries.

4.7.1 Equipment required

Supplier: Lovibond Tintometer Ltd.
Kit description: Lovibond 2000 Comparator, Hazen scale disc NSA (Range 5 – 70 Hazen Units), Nessler tubes AF306, Nessler attachment DB412 (see illustration).
Cost: £58.00.

Supplier: Camlab Ltd.
Kit description: HACH high range colour test kit Model CO – 1.
Range: 0 – 100 colour units.
Deionised water.
Cost: £51.00.

Procedure:
1. Fill up a clean comparator tube with your sample and compare the colour with a matched comparator tube containing deionised water in the comparator using the Hazen colour disc.
2. Record the colour reading from the disc.
3. Measure the pH of the sample and record at the same time. (See notes on pH measurement).
4. If turbidity is present report as apparent colour.

111

5. If the colour exceeds the disc reading dilute the sample with deionised water until the colour is within the range of standards.
6. Rinse out the tubes and place them carefully in their storage position.

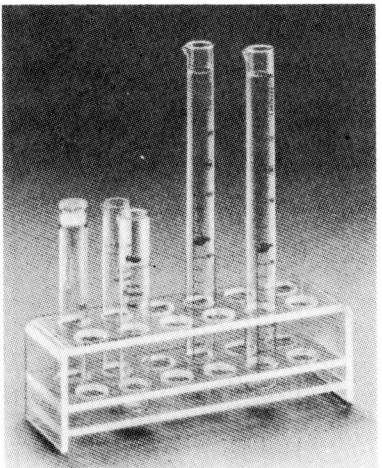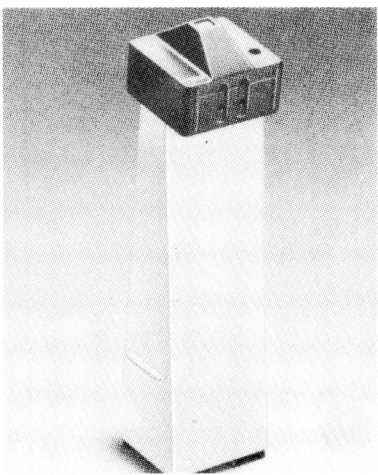

Figure 4.9   Lovibond Nessler tubes & DB412 viewer

5. Concluding remarks

There are many test kits available for water testing and more are being developed. Some of the equipment described in this report is expensive and the cost may only be justified by the accuracy required, the frequency of the determination and its relevance to the user.

It is very important not to waste money on determinations which are not required. As far as possible the user should choose equipment which can be used for more than one application. For example, one colour comparator may utilise several discs for different determinations. Manufacturers will sell components, refills and discs separately allowing the user to build up his range of kits. If a specific ion meter is purchased the additional purchase of other electrodes will make the basic meter more cost effective. One specific ion meter can be used with several electrodes: pH, fluoride, nitrate, dissolved oxygen and chloride. The digital titrator is a good example of one instrument with many functions bringing titrations effectively into field chemistry.

There are a few manufacturers who prepare chemical testing kits to a customer's specification, or have packaged kits for several determinations. They are HACH, pHOX, Lovibond, Lamotte, Wilkinson and Simpson, and they should be contacted for further details.

There is a need for improvements in field testing techniques and equipment. For example a cheap portable incubator capable of operating at 35 °C and 44.5 °C for about 10 petri dishes is urgently needed. Cheaper, more accurate chemical field tests will always be required, researchers and manufacturers please note! Better faecal indicator organisms for tropical applications and media for counting them must be researched.

'Good field workers must cooperate' is the message on a Tanzanian brochure for health workers. The public health engineer in the past has effected great social revolutions by providing sanitation and good water supply to the general public in Europe. If progress is to be made in developing countries their people (their greatest resource) must be healthy. That health can be provided by clean water and good sanitation, and public health engineers working effectively in developing countries.

It is hoped that this book will serve as an information source and assist field testing of water in developing countries. It is important that users of field equipment for water testing point out their experiences to the author and exchange their findings so that progress may be made in assessing the performance of kits under arduous conditions. Inevitably in producing a review or 'state of the art' type of book something is missed, neglected or overlooked so please regard the book as a basis for discussion and exchange of views.

Acknowledgements

The kind permission of the following organisations to reproduce material in this book is gratefully acknowledged:
HACH Chemical Co Ltd
Millipore (UK) Ltd
Wilkinson and Simpson Ltd
Kent-EIL Ltd
Lovibond-Tintometer Ltd
E. Merck

I would like to thank the staff of the Water Research Centre for their encouragement, comments and assistance during the production of this publication. I am also grateful for assistance from Sue Ball (artwork), K. R. Pugh (audio-visual services) Margaret Ince and Sue Hutton (editing), Eileen Kearins and Janet Smith for their efficient typing under pressure and finally to my wife Susan and daughter Rosemary for constant help, support and love.

References

APHA 1980
   Standard Methods for the Examination of Water and Waste Water. 15th Edn, New York, APHA/AWWA/WPCF. ISBN 0875530915.
AWWA 1978
   Simplified Procedures for Water Examination. AWWA Manual M12 (including supplement). Denver, Colorado. American Water Works Association.
BARRELL, R A E and ROWLAND M G M 1979
   The relationship between rainfall and well water pollution in a West African (Gambian) village. Journal of Hygiene 83, 143 – 150.
CAIRNCROSS, S and FEACHEM R 1978
   Small Water Supplies, Bulletin No 10 London Ross Institute. ISBN 0900995106.
COOK, J M and MILES, D L 1980
   Methods for the Chemical Analysis of Groundwater. Institute of Geological Sciences. Report 80/5. London, HMSO. ISBN 0118841831.
COX, C R 1969
   Operation and Control of Water Treatment Processes, Monograph Series No. 49, 3rd impression 1973 Geneva, WHO.
FEACHEM, R G, BRADLEY, D J, GARELICK, H and MARA, D D
   Sanitation and Disease: Health Aspects of Excreta and Waste Water Management. Baltimore, Ohio. John Hopkins, University Press.
GORCHEV, H G and OZOLINS, G 1982
   WHO Guidelines for Drinking Water Quality Proc. I.W.S.A. Congress, Sept. 1982 Zurich.
HACH, C C 1979
   Introduction to Turbidity Measurement, Technical Information Series Booklet No 1, 2nd Ed. Loveland, Colorado, Hach Chemical Co.
HACH, C C 1980
   Hach Products for Water and Waste Water Analysis. Cat No 12A, Loveland, Colorado. Hach Chemical Co.
HMSO 1969
   The Bacteriological Examination of Water Supplies, Reports on Public Health and Medical Subjects. Report No 71. Dept of Health and Social Security. Welsh Office, Ministry of Housing and Local Government, London, HMSO.

HMSO 1979
The Measurement of Electrical Conductivity and the Laboratory Determination of the pH Value of Natural Treated and Waste Waters, 1978. Methods for the Examination of Waters and Associated Materials. London, HMSO.

HOLDEN, W (Editor) 1970
Water Treatment and Examination. London. J & A Churchill. ISBN 07000147373.

HUISMAN, L, DE AZEVEDO NETTO, J M, SUNDARESAN, B B, LANOIX, J N and HOFKES, E H (Editor) 1981
Small Community Water Supplies: Technology of Small Water Supply Systems in Developing Countries, Technical Paper 18, International Reference Centre for Community Water Supply and Sanitation, Rijswijk, The Netherlands.

HUTTON, L G and LEWIS, W J 1980
Nitrate Pollution of Groundwater in Botswana. Proc 6th WEDC Conf Zaria. Nigeria. Loughborough, WEDC.

IDRC 1981
Rural Water Supply in China. Ottawa, Out IRDC-TS25e. ISBN 0889362610.

INSTITUTE OF HEALTH, CHINA 1981
Instructions of Portable Drinking Water Chemical Test Kit. Dept of Environmental Health. Institute of Health Chinese Academic of Medical Sciences. Beijing.

INTERNATIONAL LABORATORY 1981
Buyers Guide 1981. Vol 10, No 9 International Laboratory.

JOHNSON DIVISION UOP 1975
Groundwater and Wells, 4th Printing, St Paul Minnesota, Johnson UOP Inc.

LA MOTTE CHEMICAL
Water Analysis Directory 2nd Ed. Code 1588. Chestertown, Maryland. La Motte Chemical Products Company.

LEWIS, W J, FOSTER, S S D and DRASER, B S 1982
The Risk of Groundwater Pollution by on-site Sanitation in Developing Countries – a Literature Review – IRCWD Report No 01/82. Dubendorf Switzerland International Reference Centre for Wastes Disposal.

LOVIBOND
Colorimetric Chemical Analysis Methods, Apparatus Reagents Brochure L213 – E Salisbury, The Tintometer Ltd.

MACHERY-NAGEL 1978
pH Indicator Paper, Test Papers. Visocolor Test Kits for the Analysis of Water and Waste Water, June 1978. Duren, W Germany, Machery-Nagel and Co. GmbH.

MARA, D D   1974
Bacteriology for Sanitary Engineers. Churchill Livingstone. ISBN 0443009805.

McJUNKIN, F E   1976
Surveillance of Drinking Water Quality, WHO Monograph Series No 63, Geneva, World Health Organization.

MERCK   1982
Merckoquant Tests for the Semi-quantitative Determination of Ions and Compounds, Darmstadt, W Germany, E Merck Aquaquant, Water Analysis System. Darmstadt, West Germany, E Merck.

MILLIPORE   1973
Biological Analysis of Water and Waste Water, Application Manual AM302. Bedford, Mass. USA. Millipore Corporation.

MILLIPORE   1974
Field Procedures in Water Microbiology, AB314 Bedford, Mass, USA Millipore Corporation.

MILLIPORE   1982
A Guide to Membrane Separation Technology, Laboratory Products Catalogue Bedford, Mass. USA. Millipore Corporation.

MINISTRY OF HEALTH   1949
Water Softening, Ministry of Health, HMSO, London.

PALIN, A T   1957
The Determination of Free and Combined Chlorine in Water by Use of Diethyl-p-phenylene diamine. Journal of American Water Works Association, 49, 873-880.

PALINTEST   1982
Palintest Newsletter, Nov 1982. Gateshead, Wilkinson-Simpson Ltd.

REID, G W   1975
Water Test Kit-I: Users Manual: Appropriate Methods of Treating Water and Waste Water in Developing Countries. USAID Washington DC.

THOMAS, L C and CHAMBERLAIN, G J   1980
Colorimetric Chemicals Analytical Methods, 9th Ed, Salisbury, The Tintometer Co.

UNESCO-WHO   1978
Water Quality Surveys: A Guide for the Collection and Interpretation of Water Quality Data. Studies and Reports in Hydrology No 23, Paris UNESCO. ISBN 9231014730.

US AIR FORCE   1959
Maintenance and Operation of Water Plants and Systems, Air Force Manual 85 – 13, Washington, DC. US Govt Printing Office.

WHO   1971
International Standards for Drinking Water, 3rd Ed. Geneva, World Health Organization.

WHO   1977
"Report of Working Group on Health Hazards from Drinking Water", London 1977 Copenhagen, World Health Organization Regional Office for Europe.

WHO   1983
WHO Guidelines for Drinking Water Quality, 3 Volumes, WHO, Geneva.

WILKINSON & SIMPSON LTD   1982
Product Information Palintest Water Testing Kits and Price Lists, Gateshead, Wilkinson & Simpson Ltd.

Appendix 1 Ancillary equipment and materials for field work

1. Distilled water

For preparation of solutions, dilutions and washing equipment a supply of distilled water is necessary. In many Moslem countries the use of distilling apparatus is controlled, and it may not be wise or practical to import stills. Sources of distilled water should be sought from garages, power stations, chemists, pharmacists, hospital and other laboratories attached to schools, universities and government establishments. The distilled water should be checked with a conductivity meter, stored in a plastic bottle and transported to the field in a wash bottle with a rubber teat over the spout.

2. Deionised water

(i) In cases where distilled water cannot be obtained the water may be deionised using ion-exchange resins. Baird and Tatlock supply a wall mounted deioniser incorporating an Elgacan throw-away can of deionising resin. It has a built-in battery energised conductivity meter for assessing water quality. The amount of deionised water obtainable from the apparatus depends on the salinity of the water and for 350 mg/l TDS it yields 26 litres of deionised water, for water of 50 mg/l TDS it yields 170 litres. It is suggested that rainwater would give even greater yields.

Cost: 332/0501 Deioniser, Elgastat tropicalised version = £67.05
332/0500/10 Cartridges, Elgacan, pack of 4   = £11.72

(ii) Camlab Ltd market deionising resin contained in a wash bottle. The user merely fills the bottle, shakes up the solution, leaves it for 2 minutes and then squeezes out deionised water from the spout. The exchange capacity is not large, again depending on the salinity of the water, but it is a very useful piece of field equipment. Fifteen litres of deionised water can be obtained with hardness of 270 mg/l $CaCO_3$ water. The resin turns yellow when it is exhausted.

Cost: CDM 100 2 bottles ion exchange pack   £15.00
CDM 150 Replacement sachets of ion exchange resin
per 6   £12.00

3. Plastic ware

Under field conditions glass has only a limited life and plastics are more practical, and in some cases cheaper than glass.

Costs:
Camlab Ltd market Azlon ware:

AZ/5305-010	10 ml measuring cylinder	£0.84
AZ/5305-025	25 ml measuring cylinder	£0.88
AZ/5305-050	50 ml measuring cylinder	£0.90
AZ/5303-100	100 ml measuring cylinder	£1.50
AZ/5305-250	250 ml measuring cylinder	£1.89
AZ/1002-010	10 ml beaker	£0.36
AZ/1002-050	50 ml beaker	£0.44
AZ/1002-100	100 ml beaker	£0.46
AZ/1002-250	250 ml beaker	£0.70
AZ/3085-125	125 ml conical flask	£2.53
AZ/3085-250	250 ml conical flask	£3.00
AZ/1421-250	Wash bottle 250 ml	£0.94
AZ/1421-500	Wash bottle 500 ml	£1.10
PPF 25 3	Plastic filter funnels 2.5 inches diameter	£0.95
PPF 45 1	Plastic filter funnels 4.5 inches diameter	£1.30
WN/1001-125	Filter Paper No. 1 (box of 100) 12.5 cm diameter	£1.10
LHW TS1	PVC Teats 2 ml Pack of 10	£4.00

4. Glassware

Bottles, preferably Pyrex or borosilicate glass for use in bacteriological testing, can be obtained from most laboratory suppliers e.g. Camlab, Baird and Tatlock, Gallenkamp. A 250 ml Pyrex bottle with screw cap will cost about £1.60.

Camlab Ltd
Dispensing bottles, Polystop with graduated pipette 1 ml JP/L 700A Capacity 135 ml will cost £1.34 each.

5. Miscellaneous items useful for field work

Aluminium foil—for bacteriological analysis.
Spare batteries for equipment.
Superglue.
Swiss Army knife (store in hold baggage on aircraft).
Adaptors for recharging apparatus.
Electrical wires and insulating tape.
Ball of string.
At least 2 adjustable spanners up to 1'' nut size.
Autoclave tape for checking autoclave performance
    (Ref PT/700/600/200 Camlab £7.16).
Masking tape for labels.
Compass (magnetic).
Marking pens (waterproof).

Box of matches. Gas torch.
Protective clothing.
First Aid kit including Snake bite kit (if possible).

6. Notes on transportation

Alcohol, ethyl and methyl, is inflammable and is best purchased locally overseas rather than being transported by air. Remember Moslem countries may prohibit the importation of ethyl alcohol and most customs authorities can be awkward and cause trouble. Butane Gas torches should never be transported by air since their explosion on board could devastate an aircraft.

Liquids, whether transported by air or surface, will leak out of their containers unless the lids are leakproof and secured.

Equipment must be protected from vibration as much as possible by foam rubber, expanded polystyrene or other means. Assume when packing all equipment that it will be dropped at least 1 metre on to hard ground and protect it accordingly.

Appendix 2
Prices and ordering

As far as possible the prices quoted for equipment and apparatus are correct at the time of production of this publication. The prices shown do NOT include Value Added Tax (VAT, currently at 15%), carriage or packing. For orders below £25.00 some companies will also charge a handling fee

Goods exported may not be subject to VAT but most companies will advise overseas buyers on this point.

It is suggested that prospective buyers obtain an up to date quotation for specific equipment and apparatus.

When sending for quotations the manufacturers can be asked to supply free literature on their products which is a useful source of information.

Appendix 3
List of Suppliers

Baird and Tatlock (London) Ltd
PO Box 1
Romford RM1 1HA
UK
Tel: 01-590 7700
Telex: 24225

Camlab (Cambridge) Ltd
Nuffield Road
Cambridge CB4 1TH
UK
Tel: 0223 62222
Telex: 817664

Lovibond-Tintometer Ltd
Waterloo Road
Salisbury SP1 2JY
UK
Tel: 0722 27242
Telex: 47372

Sartorius-Instruments Ltd
18 Avenue Road
Belmont, Sutton
Surrey
UK
Tel: 01-642 8691
Telex: 946918

BDH Chemicals Ltd
Broom Road
Poole
Dorset BH12 4NN
UK
Tel: 0202 745520
Telex: 41186

MSE-Fisons Scientific Ltd
Manor Royal
Crawley
Sussex RH10 2QQ
UK
Tel: 0293 31100
Telex: 87119

A Gallenkamp & Co Ltd
PO Box 290
Technico House
Christopher Street
London EC2P 2ET
Tel: 01-247 3211
Telex: 886041

Millipore (UK) Ltd
Millipore House
11-15 Peterborough Road
Harrow
Middlesex HA1 2YH
UK
Tel: 01-864 5499
Telex: 24191

Wilkinson & Simpson Ltd
Palintest House
57 Queensway
Team Valley Estate
Gateshead
Tyne & Wear NE11 0NS
UK
Tel: 0632 872164
Telex: 537479

George Kent—EIL Ltd
Hanworth Lane
Chertsey
Surrey KT16 9LF
UK
Tel: 0293 62671
Telex: 264022

Whatman Lab Sales
Springfield Mi-1
Maidstone
Kent ME14 2LE
UK
Tel: 0622 674821
Telex: 96113

Oxoid Ltd
Wade Road
Basingstoke
Hampshire RG24 0PW
UK
Tel: 0256 61144
Telex: 858793

NEERI
National Environmental Engineering
Research Institute
Nehru Marg
Nagpur 20
India

Prestige Group Ltd
14/18 Holborn
London EC1N 2LQ
UK

Centronic Sales Ltd
Centronic House
King Henry's Drive
New Addington
Croydon CR9 0BJ
Tel: 0689 42121
Telex: 896474

Gelman Sciences Inc.
Laboratory Diagnostics
600 South Wagner Road
Ann Arbor
Michigan 48106
USA
Tel: 313 665 0651
Telex: 810 223 6037

Nuclepore Corp
7035 Commerce Cr.
Pleasanton
California
USA
Tel: 415 462 2230
Telex: 337 751

Grant Instruments (Cambridge) Ltd
Barrington
Cambridge CB2 5QZ
Tel: 0763 60811
Telex: 81328

pHOX Systems Ltd
Ivel Road
Shefford
Bedfordshire SG17 5JU
Tel: 0462 813103
Telex: 825480

Appendix 4 List of manufacturers (Head Office)

Millipore Intertech Inc
PO Box 255
Bedford
Massachusetts 01730
USA
Tel: 617275 9200
Telex: 92 3457

E Merck
D-6100 Darmstadt
Frankfurter Str 250
Postfach
West Germany
Tel: 061 51721
Telex: 419325

Hach Chemical Co
PO Box 389
Loveland
Colorado 80539
USA
Tel: 303 669 3050
Telex: 9109309038

The La Motte Chemical Products
   Company
Box 329
Chestertown
Maryland 21670
USA
Tel: 301 778 3100

Lovibond-Tintometer Ltd
Waterloo Road,
Salisbury SP1 2JY
UK
Tel: 0722 27242
Telex: 47372

Wipex Products Ltd
17 Weymouth Mews
London W1N 3FQ
UK
Tel: 01 637 2615

Sartorius GmbH
Postfach 19
D.3400 Gottingen
West Germany
Tel: 0551 3081
Telex: 96723

Machery Nagel Co GmbH
PO Box 307
D-5160 Duren
West Germany
Tel: 02421 61071
Telex: 0833893

Wilkinson & Simpson Ltd
Palintest House
57 Queensway
Team Valley Estate
Gateshead
Tyne & Wear NE11 0NS
UK
Tel: 0632 872164
Telex: 537479

Orion Research
380 Putnam Avenue
Cambridge
Massachusetts 02139
USA
Tel: 617 864 5400
Telex: 921466

Horiba Ltd
Miyan-Ligashi
Kisshoin
Minamibu
Kyoto
Japan